Neuropsychology Laboratory Manual

Norman B. L. Ferguson

AUGSBURG COLLEGE

ALBION PUBLISHING COMPANY

San Francisco, California   94133

Acknowledgement
The rat brain sections shown in Chapters 4 and 5 were modified and redrawn from *The Rat Brain: A Stereotaxic Atlas of the Forebrain and Lower Parts of the Brainstem* by J. F. R. König and R. A. Klippel. Copyright 1963 by the Williams & Wilkins Co.

ALBION PUBLISHING COMPANY
1736 Stockton Street
San Francisco, California 94133

Copyright© 1977 by Albion Publishing Company

Printed in the United States of America

Library of Congress Catalogue Card Number 76–45272
ISBN 0 87843 630 8

Preface

This manual is a compilation of a number of different laboratory exercises that I have used over the past few years. Prior to the development of this manual such laboratory exercises were conducted with the use of dittoed handout sheets that were supplemented by brief blackboard sketches. The impetus to bring together these assorted handouts and to add the illustrations has come mainly from former students of mine. It is my hope that this manual, and especially the figures and diagrams, will prove to be very useful and informative to each student.

A notion underlying all the exercises included in this manual is that of allowing the student to attain a more complete understanding of the structural-functional relationships of the nervous system. It is my opinion that functional information regarding the nervous system can be better understood if the student has a clear idea of the structural relationships involved. This laboratory manual is designed to help the student achieve this goal.

Chapter 1 provides an overview of neuroanatomy. The major divisions of the nervous system are illustrated and a number of important terms and definitions are introduced. Much of the information in Chapter 1 is a necessary prerequisite for successfully dealing with the latter chapters. Chapter 2 details a step-by-step procedure for gross dissection of the sheep brain. Many schematic diagrams are provided to aid in the identification of structures. While doing the sheep brain dissection the student should try to develop a three-dimensional perspective of the brain. The sheep brain was chosen for the dissection because it is a fairly representative mammalian brain that is also reasonably large and readily obtainable. Chapter 3 outlines a step-by-step procedure for the dissection of a sheep eye. The visual system is the sensory system that most readily lends itself to a meaningful examination through gross dissection. The sheep eye is used because it is large and readily obtainable. Chapter 4, which provides a set of rat brain serial sections, is included for two primary reasons. First, it gives the student an opportunity to make a comparative analysis between two mammalian brains—sheep and rat. Since the rat brain is relatively small, the illustrations are presented in the form of enlarged microscopic sections that are in serial order. This will permit the student to study some of the finer structures of the brain that could not be seen during the gross dissection of the sheep brain. Secondly, the rat brain serial sections are presented as a prelude to the information on rat stereotaxic surgical technique that is given in the next chapter. Chapter 5 details a step-by-step procedure that can be used when doing subcortical brain surgery on a rat or other small rodent. To effectively analyze the results of such experimental procedures, one must have an adequate knowledge of the anatomy of the rat brain. Chapter 6 is included to provide the student with some examples of human applications that can be made of the knowledge of the brain's structural-functional relationships.

References and Suggestions for Further Reading are provided at the end of each chapter so that students can readily locate other sources of pertinent information. Students who want more detailed information and illustrations are referred to these sources. A *Glossary of Neuroanatomical Terms* is provided in Appendix A in order that students may learn the derivations and origins of many of the seemingly cryptic structural names. A better understanding of a term's origin may facilitate its being remembered.

Norman B. L. Ferguson

Minneapolis, Minnesota
August, 1976

Contents

To Janine

1

Basic Neuroanatomy

The information in this initial chapter is meant to provide a general orientation to the anatomical relationships of the nervous system. Directional terms, definitions, and planes of reference will be covered. Additionally, brief descriptions of functional relationships within both the peripheral nervous system and central nervous system are included. This chapter should provide the understanding that will be essential in dealing with the material in the succeeding chapters. Chapter 1 should also be helpful in understanding lecture and textbook material that deals with behavioral-anatomical relationships.

1. DIRECTIONAL TERMS

When we refer to various aspects of the nervous system a standard set of directional terms is used. These terms are used because they each have a precise meaning. Refer to Figure 1.1 while reading the following descriptions. The term *dorsal* is used to refer to the top or back of the nervous system. For animals that stand on all four limbs such as cats, dogs, and rats the meaning of dorsal is quite straightforward. However, for animals that stand on only two limbs such as humans the use of the term dorsal is somewhat more complex. This is due to the fact that there is a

pronounced bend in the nervous system of humans, approximately at the level of the midbrain. This bend in the nervous system has apparently come about owing to our upright posture. In humans the term dorsal refers to the side of the spinal cord nearest to our back. It also refers to the part of our brain, when viewed from above, that is nearest our skull. The term *ventral* is essentially the opposite of dorsal. It refers to the front or bottom. The ventral side of a cat is its underbelly. In humans ventral is the front or chest side.

A second pair of directional terms is *medial* and *lateral*. Medial means toward the middle or midline, whereas lateral means toward the side or away from the midline. A third pair of directional terms is *anterior* and *posterior*. Anterior means toward the head, whereas posterior means toward the tail. Other terms that have the same meaning as anterior are *rostral* and *cephalic*. Another term that has the same meaning as posterior is *caudal*. A fourth pair of directional terms is *superior* and *inferior*. Superior can mean either "dorsal to" or "rostral to," whereas inferior can mean either "ventral to" or "caudal to."

Since the brain consists of two symmetrical halves, it is convenient to have terms that permit a variety of distinctions to be made when referring to it. The term *unilateral* is used to indicate the involvement of only one half of the

LATERAL VIEW

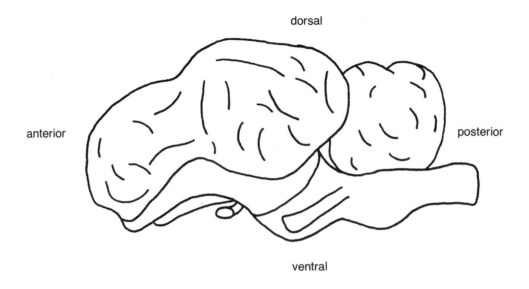

DORSAL VIEW

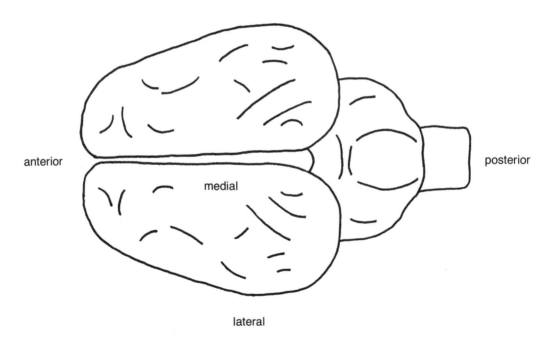

Figure 1.1. *Schematic diagrams of the brain indicating the various directional terms.*

brain (e.g., "The animal was given a unilateral lesion."). The term *bilateral* is used to indicate the involvement of both sides of the brain (e.g., "Bilateral recordings were made of the electrical activity of the hippocampus."). The term *ipsilateral* is used when we refer to the same side of the brain, whereas the term *contralateral* is used when we refer to the opposite side of the brain.

2. PLANES OF REFERENCE

When a part of the nervous system (e.g., brain) is cut so that it can be examined more closely, it is usually sectioned in one of the three planes of reference. Vertical sections that are made perpendicular to the anterior-posterior axis are referred to as *frontal* (or *coronal*) sections. This is the most frequently used plane for sectioning the brain. Sections that are made in the horizontal plane are simply referred to as *horizontal* sections. Vertical sections that are made parallel to the anterior-posterior axis are referred to as *sagittal* sections. For special purposes the brain can also be sectioned in a variety of oblique planes (see Figure 1.2).

3. STRUCTURAL TERMS AND DEFINITIONS

Following is a list of definitions of some of the most frequently used neuroanatomical terms.

a. *central nervous system* (CNS)—consists of the brain and spinal cord including nerve cell bodies and their processes, nonneural glial cells, and associated blood vessels and membranes.

b. *peripheral nervous system* (PNS)—all parts of the nervous system not included within the CNS.

c. *brain*—cells and fibers within the skull at the head end of an organism.

d. *spinal cord*—the part of the nervous system that is encased within the spinal column.

e. *neuron*—an individual nerve cell, its cell body and processes. Below is a list of the major parts of a neuron (see Figure 1.3).

   (1) *cell body* (*soma*)—the part of the nerve cell that contains the cell nucleus.

   (2) *dendrite*—a cell process that brings information into the cell.

   (3) *axon*—the primary cell process for transmitting information away from the cell body.

   (4) *axon hillock*—the junction region between the cell body and the axon; the site of spike formation.

   (5) *myelin sheath*—the lipid coat that surrounds the axon and provides electrical insulation and support.

   (6) *nodes of Ranvier*—small gaps in the myelin sheath where the cell membrane is exposed to the extracellular environment.

   (7) *terminal boutons*—enlargements on the distal ends of axons just prior to the synaptic region.

f. *synapse*—the small space between an axon terminal and another neuron where information is passed from one cell to the next.

g. *nerve*—a collection of axons, excluding cell bodies, from a large number of neurons. In the CNS called *tract, stria, bundle,* or *fasciculus.*

h. *nucleus*—a collection of neuron cell bodies. In the PNS often called a *ganglion.*

i. *gray matter*—parts of the nervous system that consist mainly of cell bodies where neuronal interaction occurs (e.g., cerebral cortex).

j. *white matter*—parts of the nervous system that consist mainly of axons or fibers (e.g., cerebral peduncles).

4. MAJOR DIVISIONS OF THE BRAIN

On the basis of embryological development the brain can be divided into a number of different sections. The three major divisions are

4

LATERAL VIEW

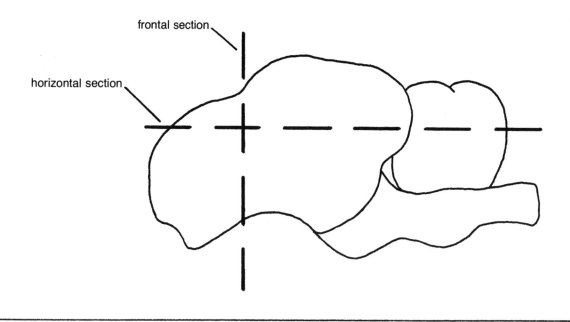

DORSAL VIEW

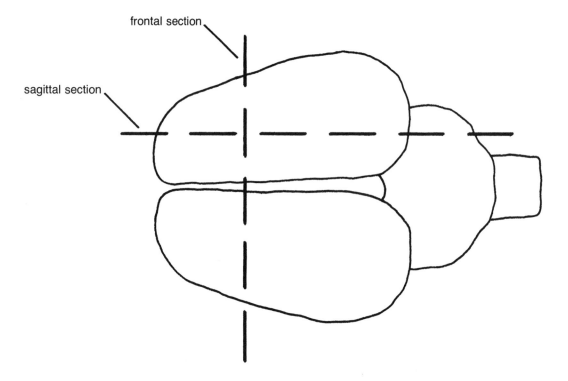

Figure 1.2. *Schematic diagrams of the brain indicating the three planes for sectioning.*

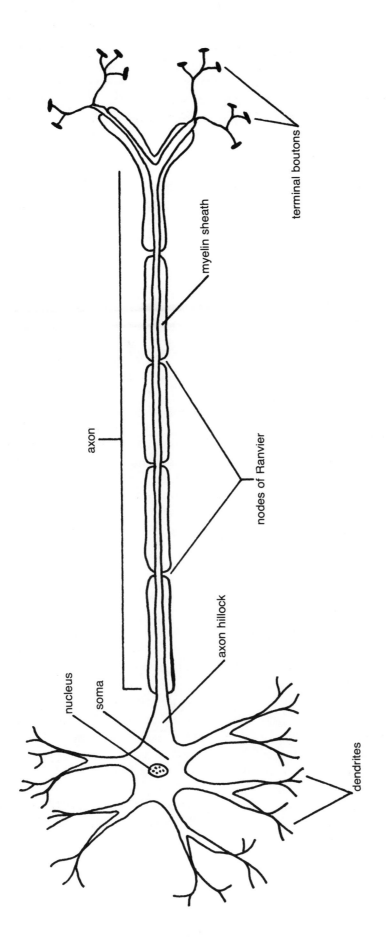

Figure 1.3. *Schematic diagram of a neuron.*

referred to as the *forebrain,* the *midbrain,* and the *hindbrain.* The forebrain can be further divided into the *telencephalon* and *diencephalon.* The hindbrain can be further divided into the *metencephalon* and *myelencephalon* (see Table 1.1).

5. PERIPHERAL NERVOUS SYSTEM (PNS)

The peripheral nervous system includes all those cell bodies and axons not included within the CNS. The PNS has two divisions generally referred to as *somatic* and *autonomic.*

a. Somatic

The somatic component of the PNS is further divided into two parts. One part consists of afferent cells and fibers that send sensory information into the CNS from the periphery. The other part of the somatic component contains the efferent fibers that innervate striated skeletal muscles. Refer to Figure 1.4 while reading the following descriptions.

Most of the nerves of the somatic component are *mixed nerves* in that they contain both sensory (afferent) and motor (efferent) axons. Motor nerve cells have their cell bodies in the ventral part of the central gray matter of the spinal cord (*ventral horn*). Axons from these cells travel out the *ventral roots* of the spinal cord, and then they enter mixed nerves that go to various body structures. Sensory nerve fibers that have their cell bodies in the *dorsal root ganglia* bring information from peripheral sensory receptors into the spinal cord. The spinal cord has ascending fibers that inform the brain about a variety of sensory inputs. In addition, the spinal cord has descending fibers that are responsible for initiating much of the activity of muscle cells.

The spinal cord has a *segmental arrangement* in that fibers innervating a specific level of the body come from a specific level of the spinal cord. A *dermatome* is the skin region innervated by one dorsal root. There is considerable overlap in the area of innervation of adjacent dermatomes. This appears to be an adaptive circumstance that permits the CNS to receive sensory information from a body region even after the primary dorsal root has ceased to function.

b. Autonomic

The autonomic component of the PNS is subdivided into two parts, the *sympathetic* and *parasympathetic* divisions. Autonomic nerves are motor in function and they innervate smooth muscles, the heart muscle, and glands. The two divisions can be distinguished on both anatomical and functional bases.

Axons from the sympathetic division leave the spinal cord at the thoracic and lumbar regions and synapse in the *sympathetic chain* that is parallel to the cord. A second axon leaves the sympathetic chain and innervates the target organ (see Figure 1.5). Activity in sympathetic nerves generally leads to the mobilization of body energy. Constriction of certain arteries, increase in heart rate, inhibition of stomach contractions, and pupillary dilation are all consequences of sympathetic activity.

Axons from the parasympathetic division leave the brainstem or sacral region of the spinal cord and travel out to within close proximity of the target organ. Here they synapse in a neural plexus onto a second neuron that innervates the target organ (see Figure 1.5). Activity in parasympathetic nerves generally leads to a conservation and storing of body energy. Dilation of certain arteries, inhibition of heart rate, increase in stomach contractions, and pupillary constriction are all consequences of parasympathetic activity. The functions of the two divisions of autonomic nerves have often been referred to as *antagonistic.*

6. FUNCTIONAL ORGANIZATION OF THE CENTRAL NERVOUS SYSTEM (CNS)

The following brief descriptions of the functions of the various portions of the CNS will be presented in sequence beginning with the

Major Divisions of the Brain

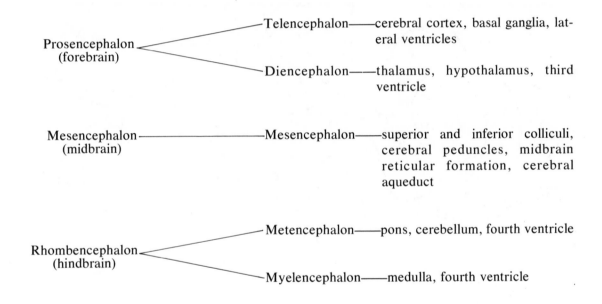

Table 1.1. *Schematic diagram of the major divisions and subdivisions of the brain including representative structures.*

DORSAL

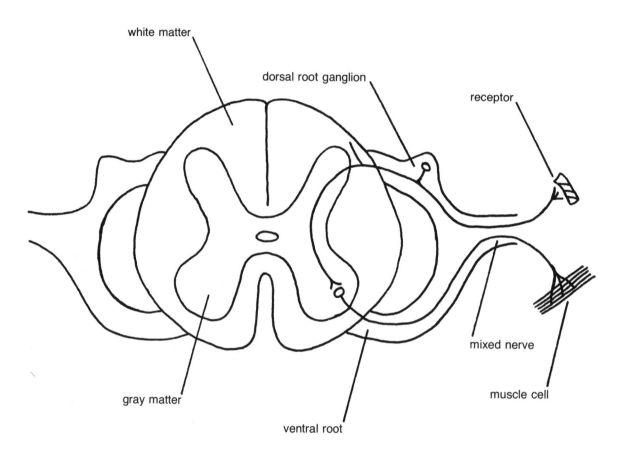

white matter

dorsal root ganglion

receptor

gray matter

ventral root

mixed nerve

muscle cell

VENTRAL

Figure 1.4. *Schematic diagram of the anatomical relationship of somatic nerves of the PNS to the spinal cord.*

9

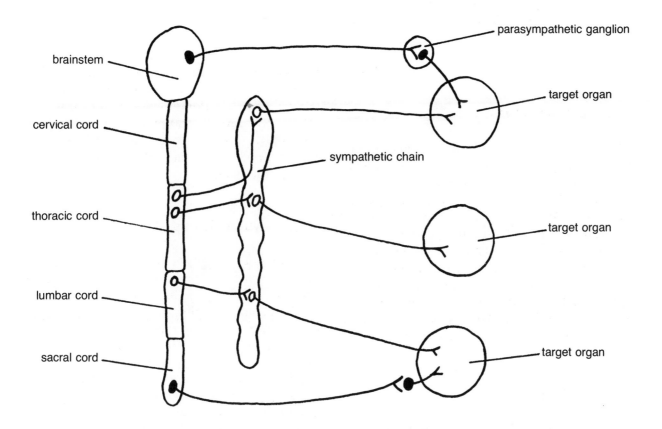

Figure 1.5. *Schematic diagram of the anatomical relationships in the sympa-thetic and parasympathetic divisions of the autonomic nervous system.*

most caudal part, the spinal cord, and progressing to the most rostral part, the cerebral cortex.

a. Spinal cord

This portion of the CNS mediates *spinal reflexes*—skeletal muscle and autonomic responses to bodily and environmental stimuli. Such reflexes persist even after the spinal cord has been separated from the brain. The cord also serves as a route of passage for ascending sensory and descending motor messages.

b. Cranial nerves

This group of nerves handles sensory and motor functions of the head. They are similar to spinal nerves, but they enter and leave the brain directly instead of traveling through the spinal cord. Most cranial nerves come from the region of the medulla and pons. There are 12 pairs of cranial nerves, and they are numbered consecutively from anterior to posterior. Some have only a sensory function, others only a motor function, and a few have both sensory and motor functions. Their specific locations and functions are given in Chapter 2.

c. Medulla

This portion of the CNS contains autonomic nerve nuclei that regulate respiration, heart rate, and gastrointestinal functions. Many cranial nerve nuclei are located in the medulla. It is also an area of passage for many ascending and descending fiber tracts.

d. Pons

This portion of the CNS contains the transverse fibers that connect one side of the cerebellum to the other. Cranial nerve nuclei and relays in the auditory system are located at the level of the pons. It is also an area of passage for ascending and descending fiber tracts.

e. Cerebellum

This portion of the CNS is located directly dorsal to the pons and functions in the regulation of motor coordination. It receives inputs from the vestibular system, spinal sensory fibers, and cerebral cortex; and it sends outputs to the thalamus and spinal cord.

f. Midbrain

This area can be divided into a dorsal part (tectum) and a ventral part (tegmentum). The dorsal part contains the superior and inferior colliculi that function in vision and hearing, respectively. The ventral part contains the third and fourth cranial nerve nuclei and the large cerebral peduncles that are part of the voluntary motor system.

g. Hypothalamus

This is physically a very small area, but it contains many nuclei that are critically concerned with biological functions such as eating, drinking, sex, sleeping, and temperature regulation. It is the major brain area concerned with autonomic function. The anterior hypothalamus controls parasympathetic functions and the posterior hypothalamus controls sympathetic functions. The hypothalamus is also directly connected to the pituitary, the master control gland of the endocrine system.

h. Thalamus

This area consists mainly of a large group of nuclei. Many of the nuclei are involved in the processing and relaying to the cerebral cortex of a variety of incoming sensory messages. The lateral geniculate body processes visual information, the medial geniculate processes auditory information, and the ventrobasal complex processes somatosensory information. Other thalamic nuclei project to association areas of the cerebral cortex, connect to the reticular formation, or form part of the limbic system.

i. Basal ganglia

These structures consist of the caudate nucleus and the lentiform nucleus. They appear to be involved in motor capacities. Dysfunctions in the basal ganglia are often involved in Parkinson's disease.

j. Cerebral cortex

This area has shown the greatest enlargement across phylogeny, particularly the neocortex. Of the total number of neurons that make up the human brain, almost three fourths of them are in the cerebral cortex. The cerebral cortex has specific sensory projection areas, and it is the origin of the motor fibers that descend through the brain and into the spinal cord. Other areas of association cortex function in more complex processes such as abstract thinking, learning, and memory. The cerebral cortex can be differentiated into a number of types on a phylogenetic basis.

REFERENCES AND SUGGESTIONS FOR FURTHER READING

Isaacson, R. L., Douglas, R. J., Lubar, J. F., and Schmaltz, L. W. *A Primer of Physiological Psychology*. New York: Harper and Row, 1971. pp. 59-89.

Leukel, F. *Introduction to Physiological Psychology, 3rd Ed*. St. Louis: Mosby, 1976. pp. 85-107.

Schneider, A. M., and Tarshis, B. *An Introduction to Physiological Psychology*. New York: Random House, 1975. pp. 73-119.

Schwartz, M. *Physiological Psychology*. New York: Appleton-Century-Crofts, 1973. pp. 1-11.

2

Dissection of the Sheep Brain

The directions that follow can be used to study any of the larger mammalian brains (dog, cat, sheep, man). Sheep brains, fixed in a 10% formalin solution, will be supplied to you for the following exercises. Be careful to avoid getting any of the formalin solution into your eyes since considerable irritation will result. If this does occur, bathe your eyes with tap water.

The following set of instruments will be needed to carry out the dissections properly: scalpel, forceps, scissors, and wooden manicure stick (orange stick). Where some of the dissections of the sheep brain are to be made on the right side of the brain and some on the left, be sure to carry out the dissection on the side directed in order not to interfere with later procedures.

Laboratory work should be supplemented by study of textbooks and atlases, particularly for a comparison with the human nervous system. Blank pages are provided at the back of the manual so that you may record your observations and make drawings.

1. BRAIN MEMBRANES AND BLOOD VESSELS

If your specimen has the brain membranes (meninges) intact, study their structure. Identify the tough outermost membrane, the *dura*

mater (most easily seen on the brainstem). The *arachnoid* is a thin translucent membrane that lies under the dura. It is most easily seen at the anterior ends of the cerebral hemispheres where it may have a black color due to the coagulated blood entrapped in the tiny vessels that are encased by the arachnoid layer. The third membrane, the *pia mater,* is very difficult to see. It is only one cell thick and it directly covers the surface of the brain tissue. These membranes provide support and protection for the delicate brain tissue. The arachnoid layer supports blood vessels that course across the surface of the brain (see Figure 2.1).

Now examine in your specimen the blood supply to the brain. Note especially the Circle of Willis and other vessels on the ventral surface. (see Figure 2.2). The hypophysis (*pituitary*) can be removed to permit better observation of the Circle of Willis. With a scalpel cut through the infundibulum (*pituitary stalk*) which can be seen emerging from the hypothalamus between the optic chiasm and the mammillary bodies. Lateral cuts between the fifth cranial nerve and the pituitary will also be necessary to remove the pituitary completely. As can be seen, the pituitary has direct connections with the base of the brain. The pituitary has been called the "master gland" in that the hormones it secretes regulate the function of numerous other glands (e.g., gonads, adrenal glands, thyroid). Find the *basilar ar-*

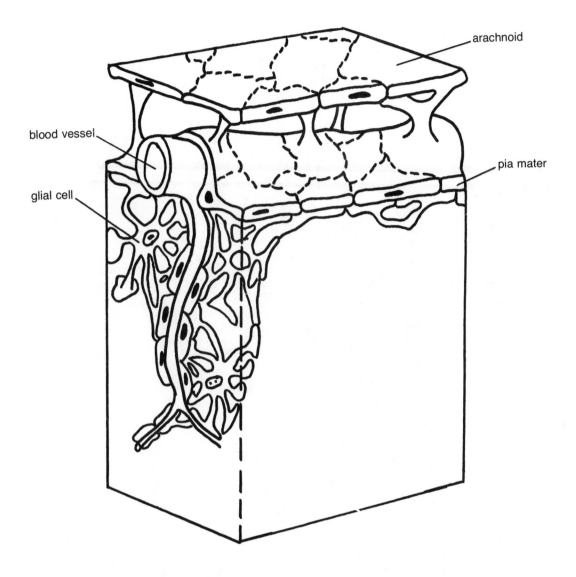

arachnoid

blood vessel

pia mater

glial cell

Figure 2.1. *Schematic diagram of a section of brain tissue showing overlying arachnoid layer and pia mater. Note position of blood vessel.*

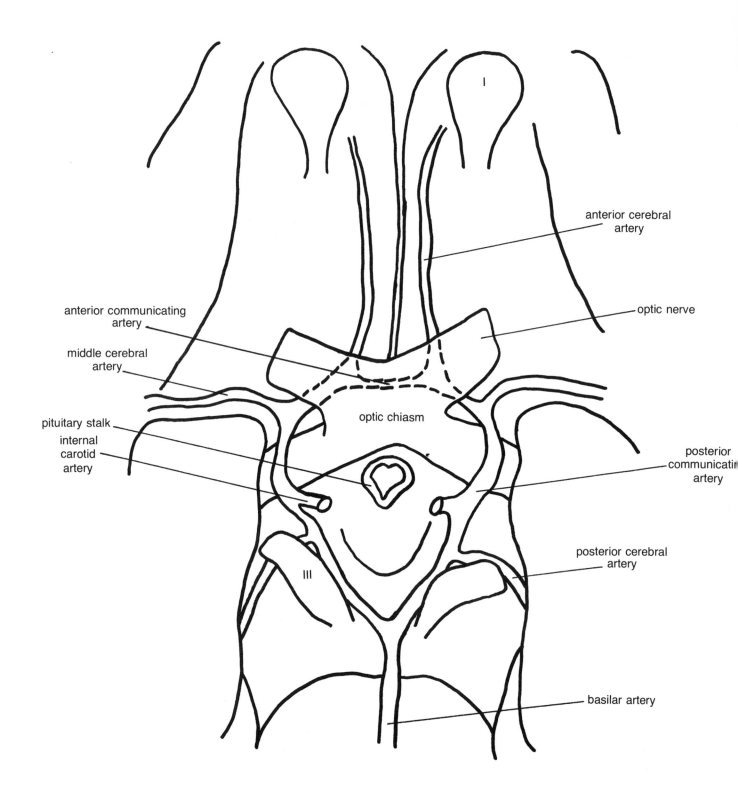

anterior cerebral
artery

optic nerve

anterior communicating
artery

middle cerebral
artery

pituitary stalk

internal
carotid
artery

optic chiasm

posterior
communicatir
artery

posterior cerebral
artery

I

III

basilar artery

Figure 2.2. *Ventral view of the brain after pituitary has been removed showing the Circle of Willis and related vessels. Note the circular pattern of the major vessels.*

tery and the *internal carotid arteries* that supply blood to the Circle of Willis; the *posterior, middle,* and *anterior cerebral arteries* that carry blood from the Circle of Willis to portions of the cerebral hemispheres; and the *posterior* and *anterior communicating arteries* that complete the Circle of Willis. Why do you suppose that the blood supply to the brain evolved in a circular form?

2. CRANIAL NERVES

Cranial nerves are bundles of axons that connect with the brain directly and do not travel through the spinal cord. There are 12 pairs of cranial nerves. Some are sensory in function, some are motor in function, and others have both sensory and motor functions. Locate the roots of the 12 pairs of cranial nerves. Each is listed below and its function is identified. The cranial nerves may be identified by either their number or name. They are numbered consecutively 1-12 from anterior to posterior.

I. *Olfactory nerve (bulb):* sensory—smell

II. *Optic nerve:* sensory—vision. The optic nerves cross at the *optic chiasm* on the midline. Posterior to the chiasm are the *optic tracts.*

III. *Oculomotor nerve:* motor—it supplies four of the muscles of the eyeball. It is large and flat and emerges from the surface of the cerebral peduncle.

IV. *Trochlear nerve:* motor—it serves the superior oblique muscle of the eyeball. This small nerve emerges from the lateral surface of the brainstem and it is often attached by connective tissue to the dorsal surface of the trigeminal nerve.

V. *Trigeminal nerve:* sensory and motor—it serves the muscles of mastication and contains sensory fibers from the face. It is behind the trochlear nerve and is very large. It emerges from the lateral border of the pons.

VI. *Abducens nerve:* motor—it serves the external rectus muscles of the eyeball. It is a small flat nerve arising from the trapezoid body.

(For cranial nerves I-VI see Figure 2.3.)

VII. *Facial nerve:* motor—it serves the muscles of the face. It is located lateral to the abducens nerve and just behind the trigeminal nerve.

VIII. *Auditory nerve (statoacoustic):* sensory—hearing and vestibular sensations. It is located just behind and slightly lateral to the facial nerve.

IX. *Glossopharyngeal nerve:* sensory—it serves the mouth and tongue area. It is located posterior and slightly lateral to the facial nerve.

X. *Vagus nerve:* sensory and motor—it serves the heart and other blood vessels and the viscera. It is located just posterior to the glossopharyngeal nerve. Cranial nerves IX and X arise as a series of small fiber bundles along the lateral border of the brainstem. They are often held close to the surface of the brainstem by connective tissue and therefore they are sometimes difficult to locate.

XI. *Spinal accessory nerve:* motor—it serves the muscles of the neck. It is large and runs along the lateral surface of the medulla.

XII. *Hypoglossal nerve:* motor—it serves the muscles of the tongue. It arises from the ventral surface of the medulla in pairs of distinct roots.

(For cranial nerves VII-XII see Figure 2.4.)

3. SURFACE ANATOMY OF THE BRAIN

Examine carefully the external form of the sheep brain. It will be necessary to remove portions of the brain membranes, especially

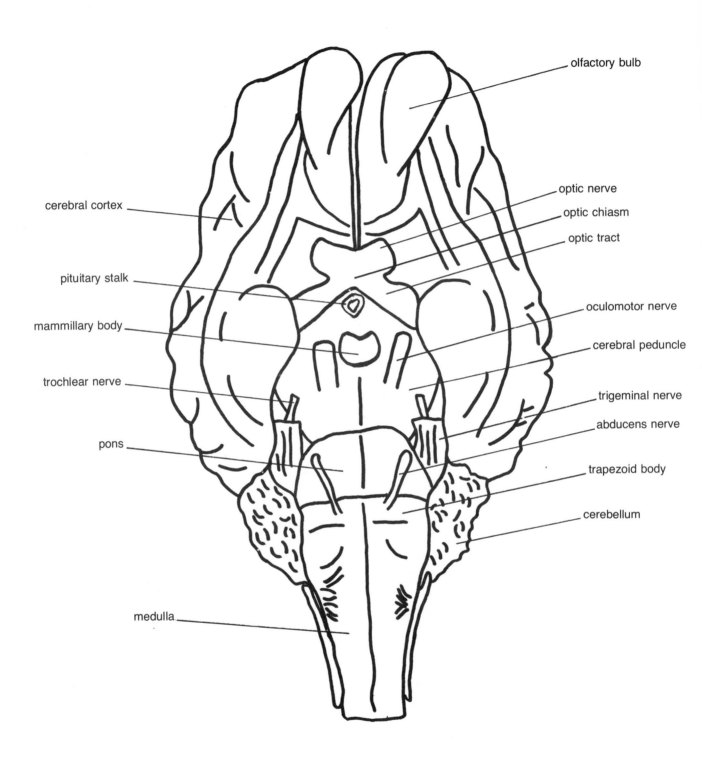

Figure 2.3. *Ventral view of the brain showing the locations of cranial nerves I through VI.*

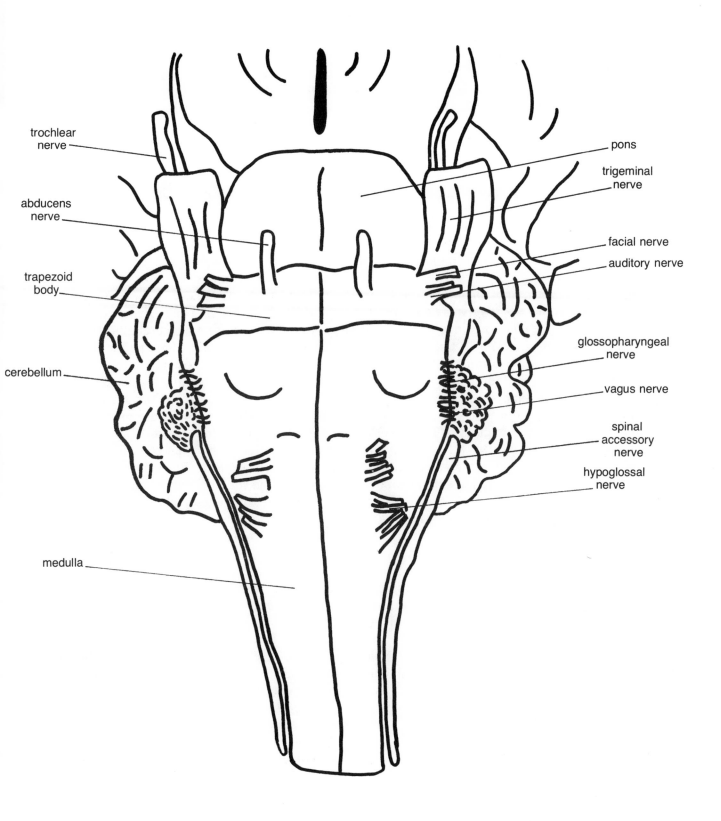

trochlear nerve

pons

abducens nerve

trigeminal nerve

trapezoid body

facial nerve

auditory nerve

cerebellum

glossopharyngeal nerve

vagus nerve

spinal accessory nerve

hypoglossal nerve

medulla

Figure 2.4. *Ventral view of the brainstem showing the locations of cranial nerves VII through XII.*

around the medulla and on the cerebral cortex, to view some of the structures. Forceps and an orange stick should be used for this purpose. Extreme care must be exercised in this procedure to avoid damaging the surface of the brain. However, it is not possible to preserve the cranial nerves entering the medulla since the membranes in this region are very thick and strong.

Identify the following structures on the ventral surface: *medulla*, *olive* and *trapezoid body* which are both synaptic areas within the auditory system, *pons, cerebral peduncles, interpeduncular nucleus* of the midbrain, *mammillary bodies, pituitary stalk, optic chiasm, hippocampal gyrus,* and *amygdala* (see Figure 2.5). Also identify the *diagonal band* and the *lateral* and *medial olfactory stria* that emanate from the olfactory bulb. The lateral olfactory stria can be traced posteriorly into the amygdala while the medial olfactory stria curves medially to the midline. The three above mentioned fiber bundles form the sides of a triangle. The area within this triangle is referred to as the *anterior perforated substance* because numerous small blood vessels enter and leave the brain in this region, giving it a stippled appearance (see Figure 2.6).

The masses of tissue that form the cerebral cortex and cerebellum are referred to as lobes or gyri (gyrus—singular) whereas the grooves between adjacent lobes are referred to as fissures or sulci (sulcus—singular). On the cerebellum identify the *vermis, hemispheres,* and *flocculus.* On the cerebral hemispheres identify the *longitudinal fissure* (see Figure 2.7). Also identify the *lateral fissure* and the *rhinal fissure* on the lateral surface of the cerebral hemisphere (see Figure 2.8).

4. CEREBRAL CORTEX AND CEREBELLAR CORTEX

Compare the cerebral cortex and the cerebellar cortex. Note especially the difference in the number of convolutions in each structure. Use a scalpel to cut off a slice about 1 cm thick from the posterior pole of the *left* cerebral cortex and a similar slice from the *left* lateral border of the cerebellum. Compare the cut surfaces and observe the relationship of the gray matter to the underlying white matter. Note the pathways of the *arcuate fibers* which loop around from one lobe of tissue to the next. Arcuate fibers have their cell bodies located in one gyrus and their axon terminals synapse on cells in the adjacent gyrus (see Figure 2.9).

5. RHOMBENCEPHALON

The rhombencephalon (or hindbrain) is composed of the metencephalon (cerebellum and pons) and the myelencephalon (medulla oblongata). Note the attachments of the cerebellum to the medulla. Cut these attachments (*cerebellar peduncles*) on each side with the aid of a scalpel. Remove the cerebellum and put it aside for future study. These peduncles should be cut as high up as possible: cut into the substance of the cerebellum if necessary rather than into the structure of the medulla. Take care not to injure the delicate membranes beneath the cerebellum that form the roof of the fourth ventricle (see Figure 2.10).

The *fourth ventricle* is sometimes called the cavity of the rhombencephalon. It is one of the major interconnected chambers within the CNS that contains cerebrospinal fluid. Posteriorly the fourth ventricle narrows to become the *central canal* of the spinal cord. Anteriorly it again narrows to become the *cerebral aqueduct* of the midbrain. Later in the dissection the *third ventricle* of the diencephalon and the *lateral ventricles* of the telencephalon will be examined.

The cerebellum forms the roof of the fourth ventricle only for a short extent between the cerebellar peduncles. Anteriorly the roof is formed by a thin sheet of tissue, the *anterior medullary velum.* Posteriorly, the roof is formed by a thin membrane, the *posterior medullary velum,* part of which is highly vascularized and folded. This is the *choroid plexus of the fourth ventricle,* composed of capillaries and epithelial tissue. The choroid plexus forms a barrier between the blood and the cerebrospinal fluid, and it is thought to be the source of the cerebrospinal fluid. The choroid plexus sits in a wide expansion of the fourth ventricle, the *lateral recess* (see Figure 2.11).

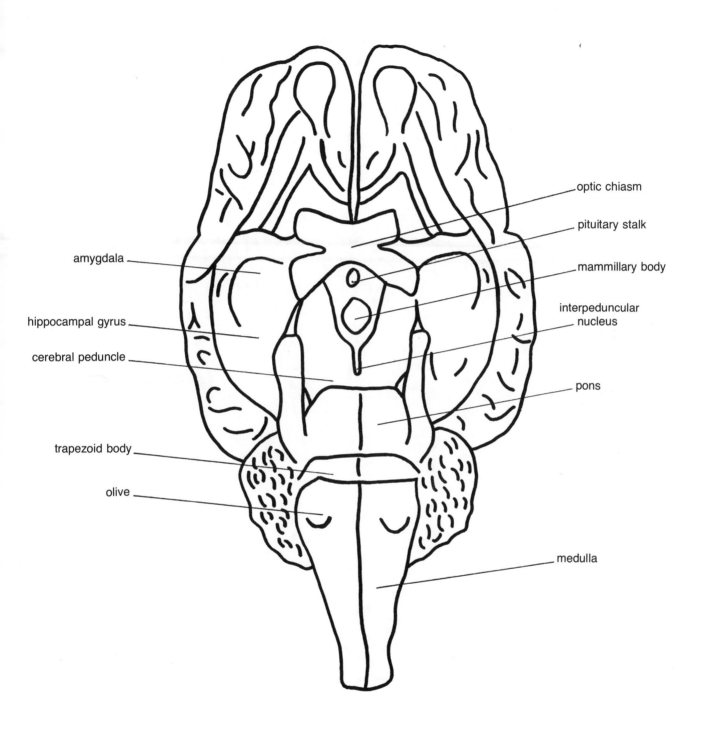

optic chiasm

pituitary stalk

mammillary body

interpeduncular nucleus

amygdala

hippocampal gyrus

cerebral peduncle

pons

trapezoid body

olive

medulla

Figure 2.5. *Ventral view of the brain showing the major features of surface anatomy.*

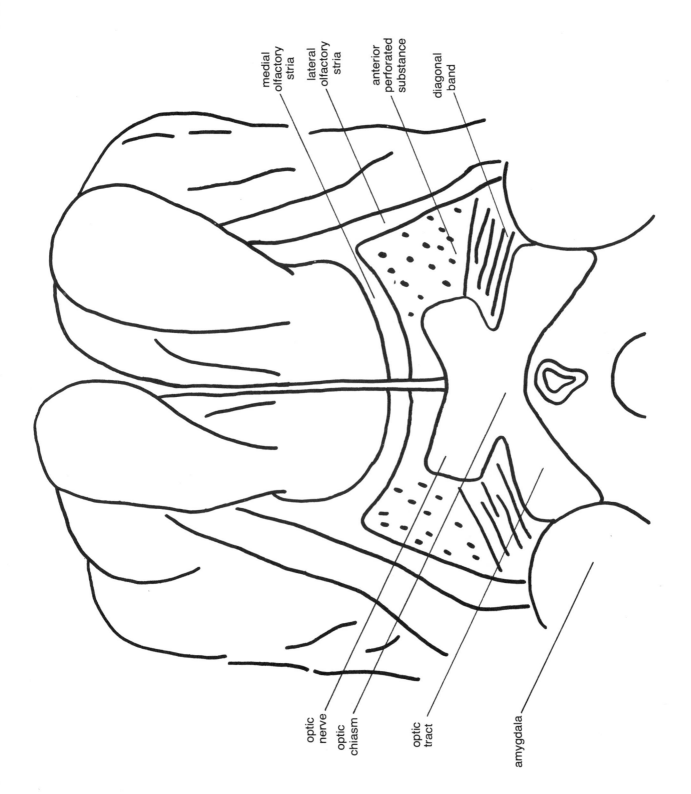

medial
olfactory
stria

lateral
olfactory
stria

anterior
perforated
substance

diagonal
band

optic
nerve

optic
chiasm

optic
tract

amygdala

Figure 2.6. *Ventral view of the anterior portion of the brain showing the major features of surface anatomy.*

21

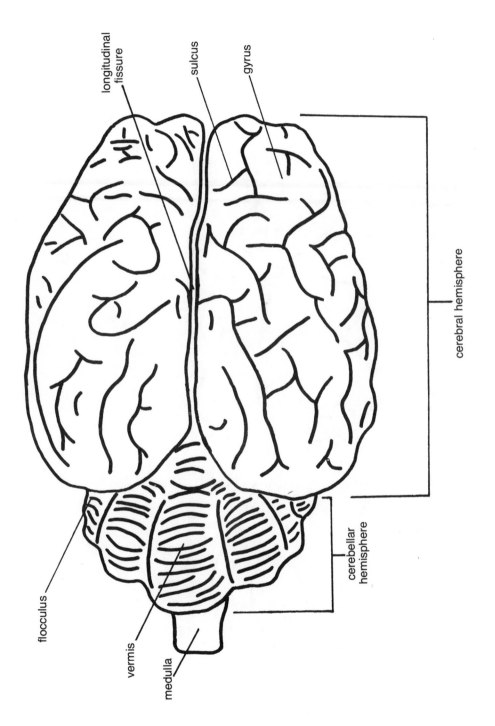

longitudinal fissure

sulcus

gyrus

cerebral hemisphere

cerebellar hemisphere

flocculus

vermis

medulla

Figure 2.7. *Dorsal view of the brain showing the cerebral hemispheres (divided into gyri or lobes by sulci or fissures and separated by the longitudinal fissure), cerebellar hemispheres, vermis, flocculus, and medulla.*

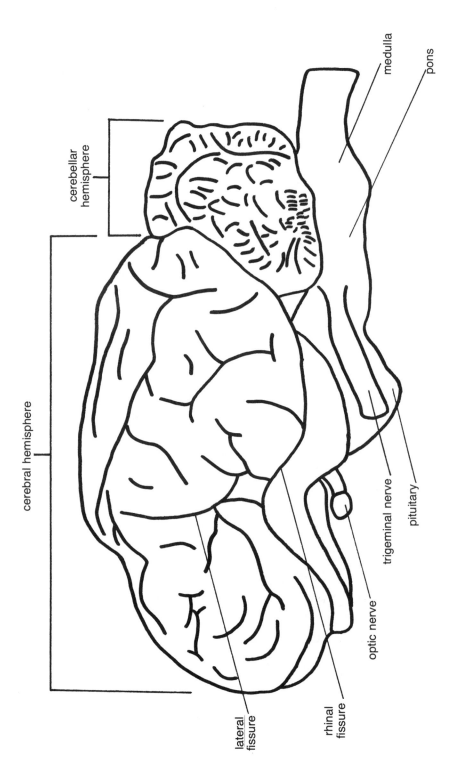

cerebellar
hemisphere

cerebral hemisphere

medulla

pons

optic nerve

trigeminal nerve

pituitary

lateral
fissure

rhinal
fissure

Figure 2.8. *Lateral view of the brain from the left side showing the left cerebral hemisphere, cerebellum, and brainstem.*

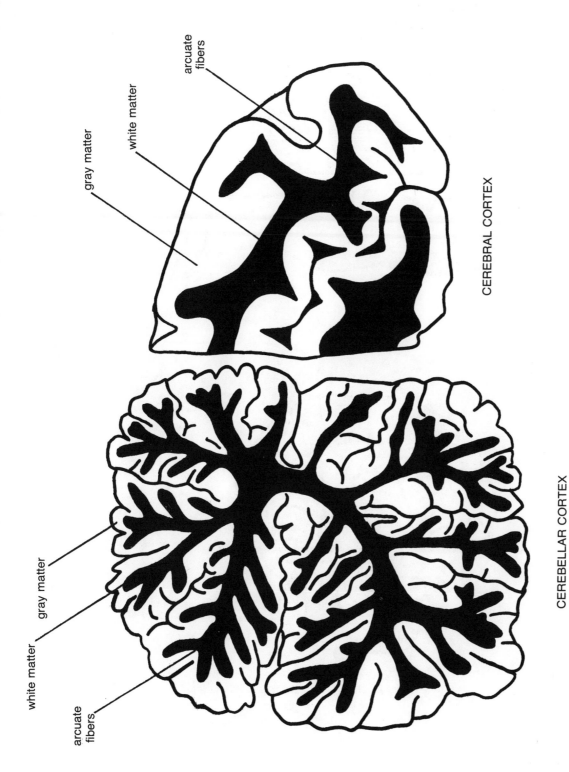

arcuate
fibers

white matter

gray matter

CEREBRAL CORTEX

white matter

gray matter

arcuate
fibers

CEREBELLAR CORTEX

Figure 2.9. *Comparison of cut sections of cerebral cortex and cerebellar cortex. Note the finer branching pattern in the cerebellar cortex.*

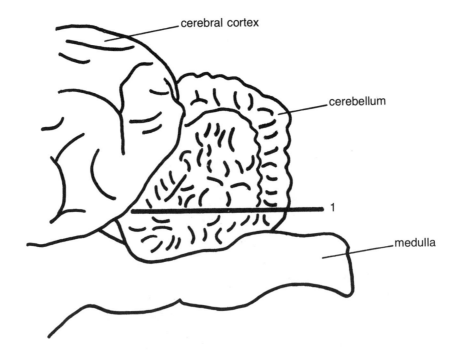

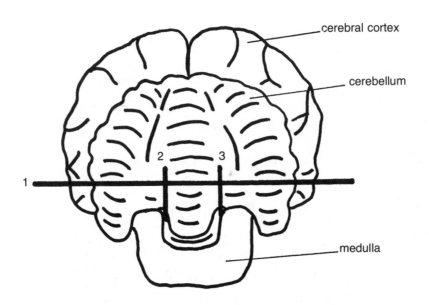

1, 2, 3=order in which knife cuts should be made

Figure 2.10. *Lateral and posterior views of the cerebellum showing where knife cuts should be made to remove the cerebellum from the brainstem.*

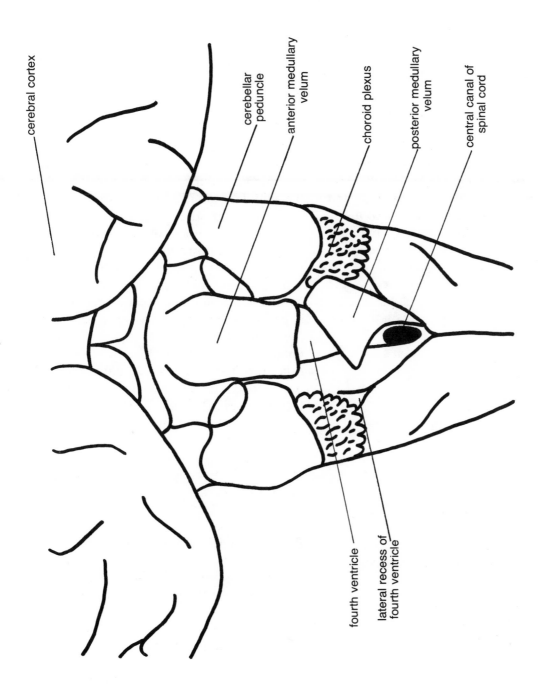

cerebral cortex

cerebellar peduncle

anterior medullary velum

choroid plexus

posterior medullary velum

central canal of spinal cord

fourth ventricle

lateral recess of fourth ventricle

Figure 2.11. *Dorsal view of the rhombencephalon with the cerebellum removed. The posterior medullary velum has been partially detached to expose the opening of the fourth ventricle.*

6. MID-SAGITTAL SECTION

Now cut the entire brain of the sheep into right and left halves. A long thin knife or a microtome blade is best suited for this purpose. The incision should pass downward through the longitudinal fissure between the cerebral hemispheres to cut through the corpus callosum in the floor of this fissure, and then through the entire brainstem. Care should be taken to make this cut smoothly and exactly in the median plane. It should be made with a single long sweep of the knife. Examine carefully the cut surfaces of the brain and identify the structures brought into view. All the prominent midline structures should be readily identifiable (see Figure 2.12). In particular the major commissures can be located. A *commissure* is a fiber bundle (white matter) that connects a structure on one side of the brain with the same structure on the other side of the brain. Thus all commissures must cross the midline.

Locate the following structures on the midline:

    medulla
    fourth ventricle
    pons
    cerebral aqueduct of the midbrain
    inferior colliculus
    superior colliculus
    pineal body
    pituitary recess of the third ventricle
    optic chiasm
    lamina terminalis
    anterior commissure
    body of the fornix
    hippocampal commissure
    septum pellucidum
    corpus callosum (genu, body, splenium)
    posterior commissure
    third ventricle
    habenula
    habenular commissure
    massa intermedia of thalamus
    stria medullaris
    mammillary body

These structures are shown in Figure 2.13.

7. CEREBELLUM

There are three cerebellar peduncles: *brachium conjunctivum, brachium pontis,* and *restiform body.* These three structures are also referred to as the superior, middle, and inferior cerebellar peduncles respectively. Examine their cut surfaces on the dorsal aspect of the medulla. With an orange stick separate the fibers of the three peduncles from each other on the cut surface of the *left* side of the brain. Then continue the separation of the superior and middle peduncles downward. The fibers of the middle peduncle can be traced down into the pons on the ventral surface of the brainstem. The fibers of the superior peduncle can be traced down into the floor of the fourth ventricle where they decussate to the other side of the brain. The posterior portion of the peduncle forms the restiform body. The superior peduncle connects the cerebellum to the cerebral hemisphere, the middle peduncle connects one side of the cerebellum to the other through the pons, and the inferior peduncle connects the cerebellum to the spinal cord (see Figures 2.14 and 2.15).

Cut the cerebellum into two halves along the mid-sagittal plane. Note the appearance of the cerebellar gray and white matter seen in the median section of the vermis (see Figure 2.9).

8. TRACT DISSECTIONS

The dissections outlined below are to be carried out on the *right* half of the brain.

a. Association fibers

Association fibers connect different regions within one area of the brain (e.g., within the cerebral cortex). The arrangement of such fibers in the subcortical white matter can be examined by careful scraping with an orange stick. Only some of these fibers are to be studied at this time. The dissection outlined in

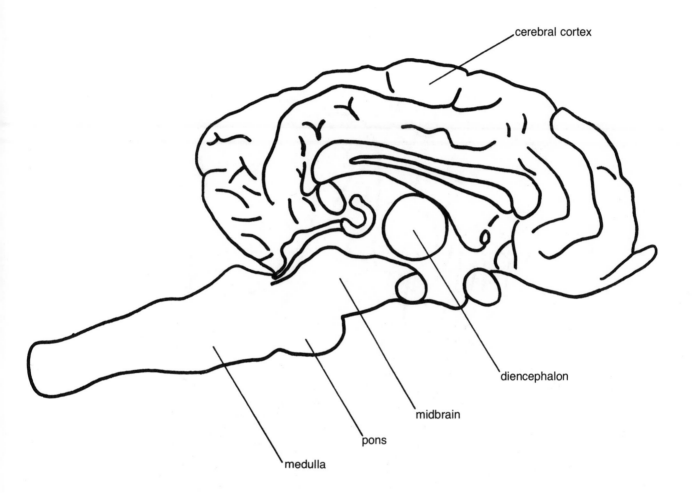

Figure 2.12. *Mid-sagittal view of the brain with the cerebellum removed.*

28

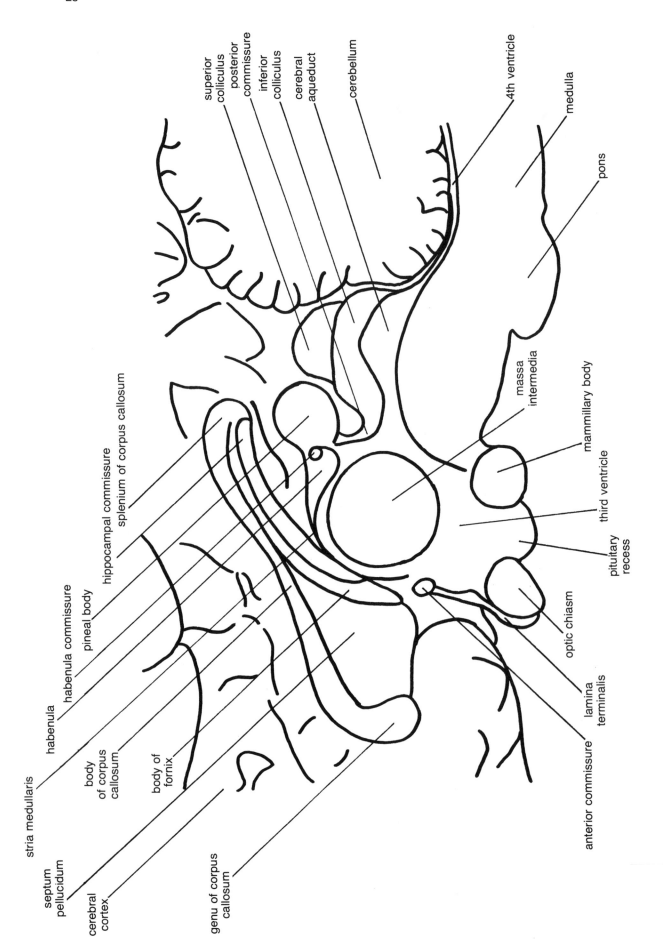

Figure 2.13. *Close-up view of mid-sagittal section showing midline structures in more detail.*

superior colliculus
posterior commissure
inferior colliculus
cerebral aqueduct
cerebellum
4th ventricle
medulla
pons
massa intermedia
mammillary body
third ventricle
pituitary recess
optic chiasm
lamina terminalis
anterior commissure
genu of corpus callosum
body of fornix
body of corpus callosum
cerebral cortex
septum pellucidum
stria medullaris
habenula
habenula commissure
pineal body
hippocampal commissure
splenium of corpus callosum

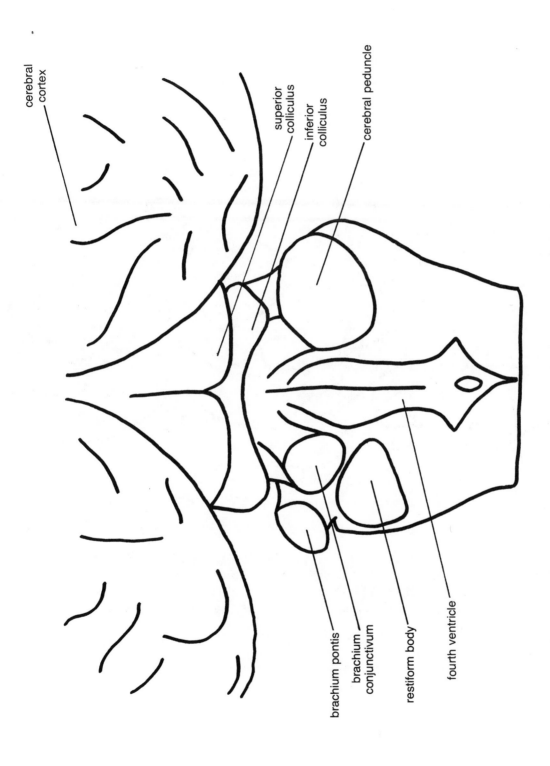

cerebral
cortex

superior
colliculus

inferior
colliculus

cerebral peduncle

brachium pontis

brachium
conjunctivum

restiform body

fourth ventricle

Figure 2.14. *Dorsal view of the rhombencephalon with the cerebellum removed. The three cerebellar peduncles have been separated on the left side of the brain.*

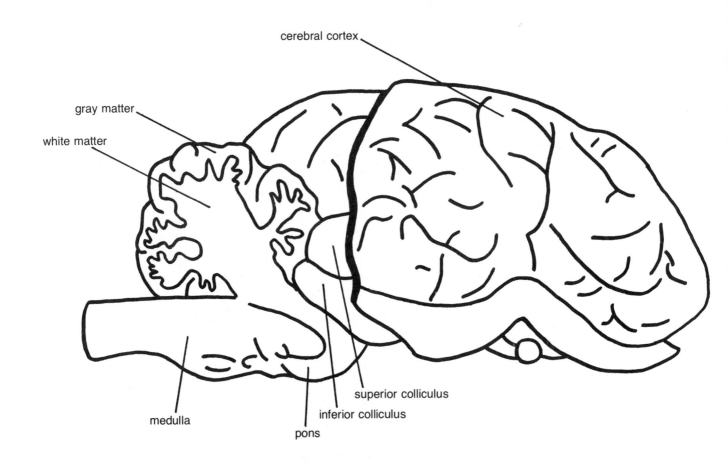

Figure 2.15. *Lateral view of the brain with the cerebellar gray matter partially dissected away and the posterior pole of the cerebral cortex removed. Note the pathway of the pons fibers extending dorsally into the cerebellum.*

this section should not be carried further than indicated.

(1) Select a region along the dorsal border of the medial surface of the hemisphere. Using an orange stick scrape away the gray matter covering two adjacent gyri to expose the *short associational (arcuate) fibers* connecting these gyri. Similar fibers lying deep in the white matter which connect the more remote gyri can be revealed by further scraping.

(2) Careful dissection of the lateral surface of the hemisphere will show that other *association fiber* systems of the cerebral cortex sweep down into the hippocampal gyrus.

(3) The *cingulum* is a long associational tract close to the medial cortical surface of the hemisphere. It runs parallel to the dorsal surface of the corpus callosum for part of its course, beginning anteriorly beneath the head of the corpus callosum. It arches upward at the genu of the callosum and passes around the splenium of the callosum posteriorly and then bends downward, forward, and laterally to the region of the hippocampal gyrus. Begin its dissection above the callosum and follow it both anteriorly and posteriorly (see Figure 2.16).

b. Commissural fibers

The fibers of the *corpus callosum* connect the neocortex of one hemisphere with that of the opposite hemisphere. Trace out a small part of these fibers to their cortical terminations (see Figure 2.16).

c. Projection fibers

Projection fibers connect structures that are at different levels within the CNS (e.g., cerebral cortex to spinal cord).

(1) *Pyramidal tract.* Locate the *pyramid* on the ventral surface of the brainstem just caudal to the pons. The pyramid is the area in which descending motor axons cross over to the opposite side of the CNS before continuing more caudally. (The pyramid is hard to see on the sheep brain owing to its mall size.) These longitudinal fibers lying just dorsal to the pons can

be exposed by stripping the pons fibers back from the cut median surface with an orange stick. By careful tracing with an orange stick the pyramidal tract can be followed spinalward as far as the pyramidal decussation. The fibers cannot be followed further. The pyramidal tract can be dissected rostrally from the pons into the cerebral peduncle (see Figure 2.16).

(2) *Corona radiata.* This name is given to those fibers running between the cerebral cortex and the diencephalon and brainstem via the internal capsule. These include the thalamo-cortical sensory projection fibers, other thalamo-cortical connections, and the cortical efferent tracts (e.g., cortico-spinal tracts). These fibers diverge from the upper border of the internal capsule like the rays of a crown: hence the name "corona radiata." The ends of the vertical corona radiata fibers should be visible through the transverse sheet of corpus callosum fibers that extends at right angles to them. The cingulum is at right angles to both of these systems.

The descending cortico-spinal motor system is one of the major projection fiber systems in the CNS. The neurons of this system have their cell bodies located in the gray matter of the motor area of the cerebral cortex. The axons of these cells descend through the hemisphere via the corona radiata and internal capsule and emerge on the ventral side of the brain in the cerebral peduncle. They continue caudally through the regions of the pons and medulla in the pyramidal tract crossing over to the opposite side of the CNS at the pyramidal decussation. After traveling down the spinal cord in the cortico-spinal tract, they synapse on ventral horn cells that directly innervate skeletal muscle cells (see Figure 2.17).

9. MESENCEPHALON AND PROSENCEPHALON

a. Rhinencephalon and limbic system

In older terminologies the rhinencephalon is often referred to as the *smell* brain. Though the olfactory bulb can be shown to have connections with the rhinencephalon, it connects

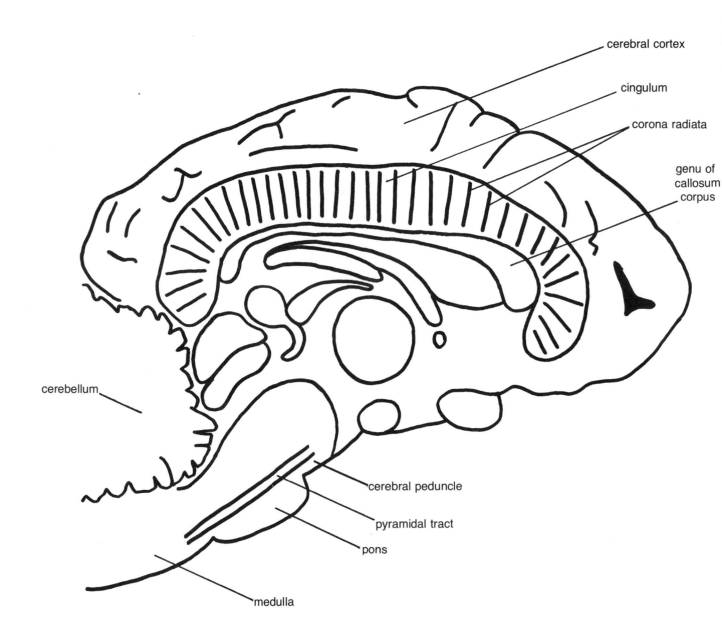

cerebral cortex

cingulum

corona radiata

genu of
callosum
corpus

cerebellum

cerebral peduncle

pyramidal tract

pons

medulla

Figure 2.16. *Mid-sagittal view of the brain showing the location of the pyramidal tract, corpus callosum, cingulum, and corona radiata. The
cingulum bundle has been scraped away to reveal the vertical
corona radiata fibers that are lateral to it.*

33

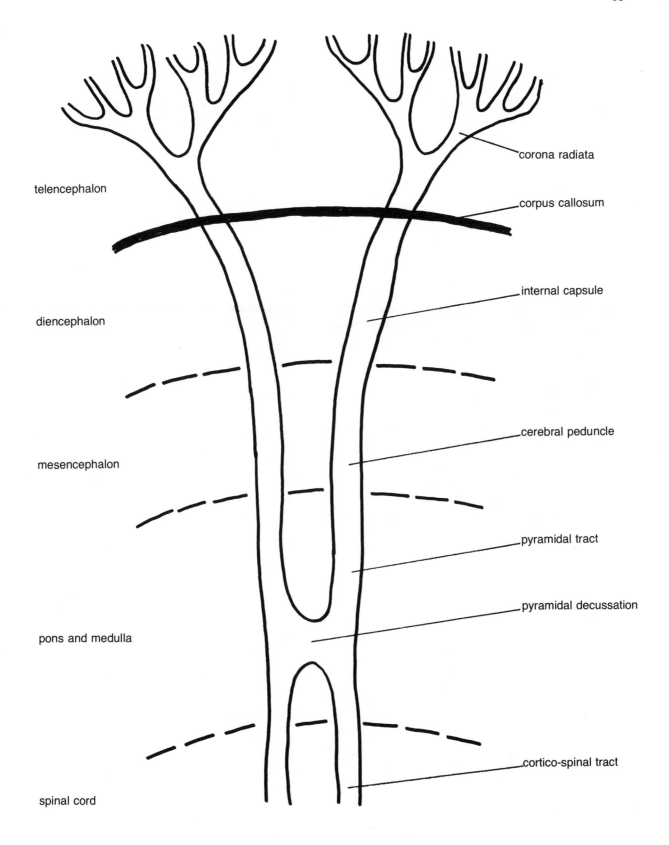

corona radiata

telencephalon

corpus callosum

internal capsule

diencephalon

cerebral peduncle

mesencephalon

pyramidal tract

pyramidal decussation

pons and medulla

cortico-spinal tract

spinal cord

Figure 2.17. *Schematic diagram of the descending cortico-spinal motor system. Major brain regions are identified.*

with only some of its parts. More recently the rhinencephalon has been divided into three interconnected systems: (1) primary olfactory structures directly related to the olfactory bulb; (2) a second system receiving fibers from the first and consisting primarily of the septal region and amygdala; (3) a third system consisting of cingulate and entorhinal cortex and the hippocampus. These appear to be remotely, if at all, related to olfactory afferents. Both the second and third systems send efferents to the hypothalamus. The *limbic system* consists of the second and third systems in addition to two diencephalic structures, the anterior thalamic nuclei and the mammillary bodies. Most of the structures of the limbic system have been shown to function in the control and mediation of a wide variety of emotional behaviors.

(1) Three fiber tracts that comprise part of the above outlined systems can be revealed through dissections carried out at the midsagittal aspect of the hemisphere. The dissections outlined below are to be carried out on the *right* half of the brain.

Fornix. Follow the column of the fornix from the body of the fornix, dissecting it out as you go, forward to a position just above the anterior commissure. Then proceed caudally and ventrally to the mammillary body. A small part turns back immediately behind the *interventricular foramen,* which connects the third ventricle to the lateral ventricle, to reach the habenula via the stria medullaris. The column of the fornix consists mainly of fibers passing out of the hippocampus by way of the *fimbria.* These fibers project mainly to the hypothalamus and subthalamus. The column of the fornix is the efferent projection tract of the hippocampus to the mammillary body and habenula (see Figure 2.18).

Mammillo-thalamic tract (tract of Vicq d' Azyr or tractus thalamo-mammillaris). This tract runs from the mammillary body forward and dorsally to the *anterior nuclei of the thalamus.* It can readily be dissected by scraping off the outer wall of the third ventricle, beginning in the region of the mammillary body (see Figure 2.18).

Habenulo-peduncular tract (fasciculus retroflexus or Meynert's bundle). This tract can also be readily dissected. It runs from the habenula into the interpeduncular nucleus between the cerebral peduncles (see Figure 2.18).

(2) Dissection of the *hippocampus.* Remove the septum pellucidum and, while looking into the *lateral ventricle,* draw apart the corpus callosum and the underlying body of the fornix. The hippocampus is located in the floor of the posterior horn of the lateral ventricle, together with the fimbria and hippocampal commissure. Now cut through the splenium of the corpus callosum to separate the fimbria and hippocampal commissure from the overlying corpus callosum. Continue this incision laterally and ventrally, cutting from the ventricular wall back into the hippocampal gyrus and cerebral cortex along the posterior and outer border of the hippocampus for its entire length down to the tip of the hippocampal gyrus (see Figure 2.19). Now, beginning in the hippocampal gyrus into which the lateral olfactory tract passes, note the shape and position of the hippocampus and its fiber tract, the fimbria, as it proceeds toward the midline (see Figures 2.20 and 2.21).

The hippocampus is a buried convolution of cortex rolled into the lateral ventricle from the ventral and posterior margins of the cerebral cortex. It is entirely covered by the hippocampal gyrus. On its ventral side the *dentate gyrus* forms another convolution that gives rise to a sheet of fibers, the fimbria, which passes forward to enter the body of the fornix. At this point some of the fibers cross to the other side of the brain in the hippocampal commissure. Other fibers descend into the diencephalon as the columns of the fornix. With the aid of an orange stick, the connections between the hippocampus and the rest of the brain can be severed and the entire hippocampus can be removed from the brain. Cutting perpendicularly into the hippocampus will reveal the spiral appearance of the hippocampal convolutions.

(3) *Anterior horn of the lateral ventricle.* Cut through the genu of the corpus callosum forward and downward toward the olfactory

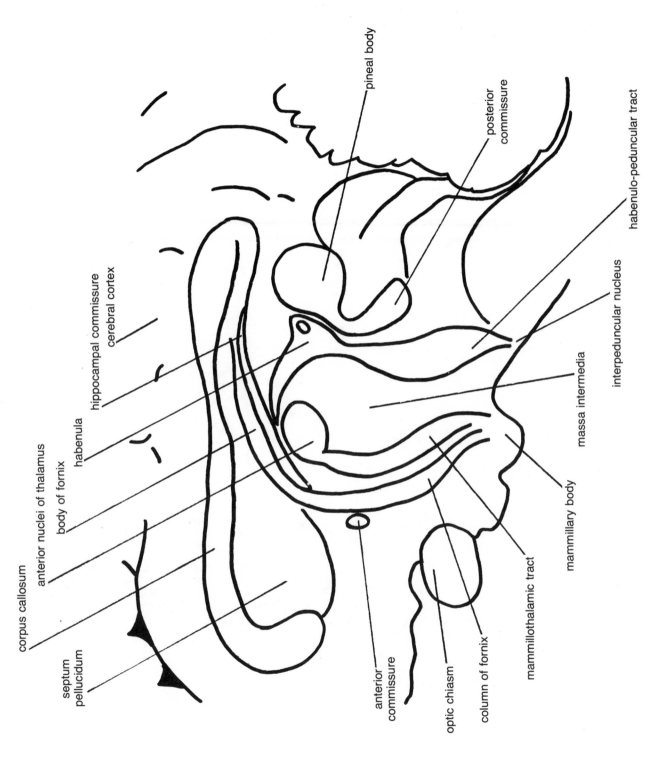

pineal body

posterior
commissure

habenulo-peduncular tract

hippocampal commissure
cerebral cortex

habenula

body of fornix

anterior nuclei of thalamus

corpus callosum

septum
pellucidum

interpeduncular nucleus

massa intermedia

mammillary body

mammillothalamic tract

column of fornix

optic chiasm

anterior
commissure

Figure 2.18. *Mid-sagittal section showing the locations of the fornix, mammillothalamic tract, and habenulo-peduncular tract.*

36

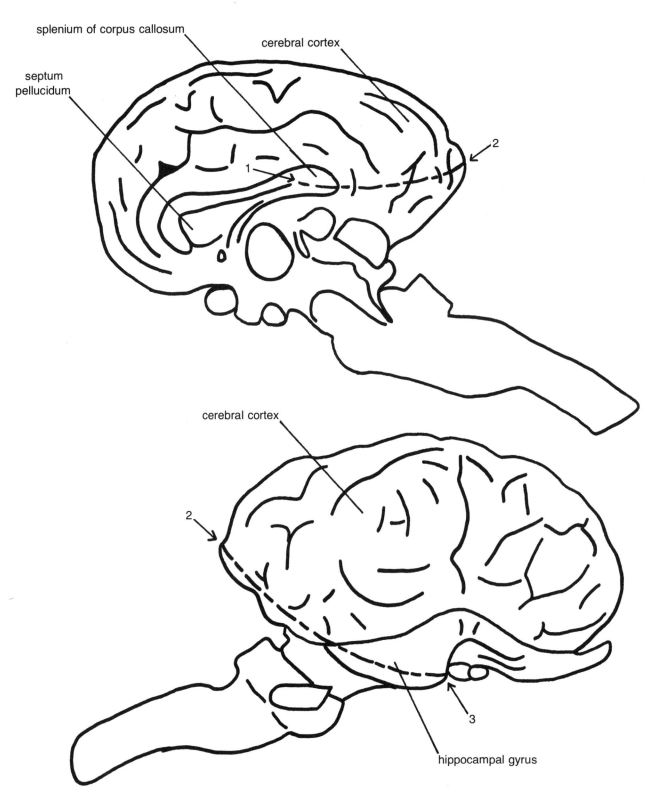

splenium of corpus callosum

cerebral cortex

septum
pellucidum

1

2

cerebral cortex

2

3

hippocampal gyrus

1, 2, 3=pathway which knife cut should follow

Figure 2.19. *Mid-sagittal and lateral views of the right half of the brain showing where to make the knife cuts that will reveal the position of the hippocampus within the lateral ventricle.*

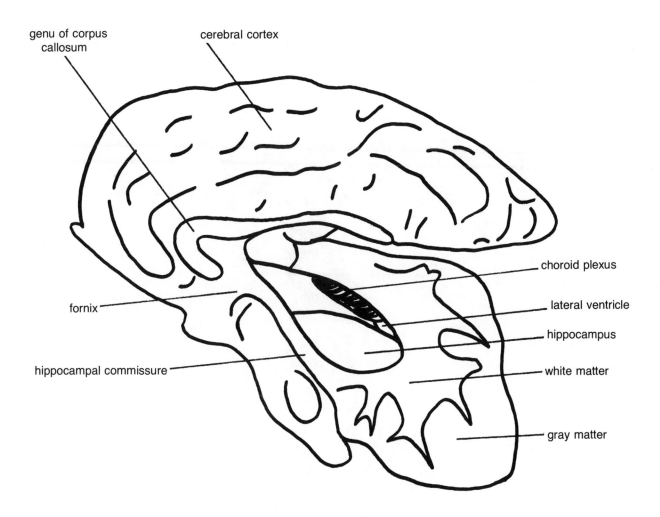

genu of corpus callosum

cerebral cortex

choroid plexus

lateral ventricle

hippocampus

white matter

gray matter

fornix

hippocampal commissure

Figure 2.20. *View of the right cerebral hemisphere after the knife cuts illustrated in Figure 2.19 have been made. The dorsal and ventral portions of the hemisphere have been spread apart to reveal the position of the hippocampus within the lateral ventricle.*

38

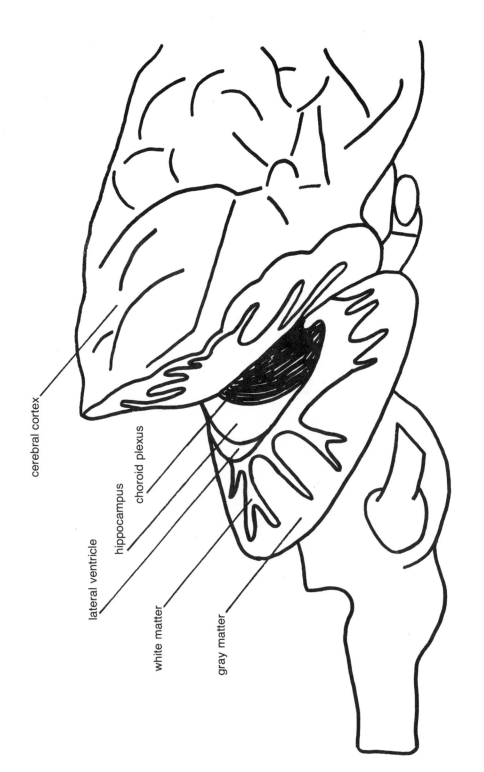

cerebral cortex

hippocampus

choroid plexus

lateral ventricle

white matter

gray matter

Figure 2.21. *Lateral view of the right side of the brain with the cerebellum and the posterior pole of the cerebral cortex removed. The hippocampus can be seen sitting in the lateral ventricle.*

bulb to open the anterior horn of the lateral ventricle. In the sheep, unlike man, this is directly continuous with the ventricle of the olfactory bulb (see Figure 2.22).

b. Other structures of the telencephalon

Beginning now in the hippocampal gyrus, note the position at the lateral border of the lateral ventricle of the *stria terminalis*. Note also the *choroid plexus of the lateral ventricle* and the *tail of the caudate nucleus*. Follow these three structures medially and anteriorly to the anterior end of the *head of the caudate nucleus*. Note the relations to the hippocampus (see Figure 2.22). The stria terminalis (which is very difficult to see in the sheep brain) runs forward into the anterior commissure and posteriorly to where it enters the *amygdala*, a small deep gray mass in the anterior tip of the hippocampal gyrus.

With the aid of a scalpel cut the *right* hemisphere into anterior and posterior sections by making a vertical cut completely through the brain just in front of the anterior commissure. Examine the cut surfaces of the hemisphere. Notice the *internal capsule* fibers passing downward and backward lateral to the caudate nucleus. The *lentiform nucleus*, a large mass of gray matter, can be seen lateral and ventral to the internal capsule. The lateral border of the lentiform nucleus is formed by the *external capsule*. Some of the fibers of the internal capsule can be traced ventrally into the cerebral peduncle and dorsally into the cerebral cortex. Notice also how the transverse fibers of the corpus callosum distribute to the cerebral cortex (see Figure 2.23).

c. Optic system

Remove the posterior lobe of the *right* cerebral cortex with a scalpel and identify the *pulvinar, lateral geniculate body,* and *superior colliculus* on the lateral surface of the thalamus and midbrain. Follow the *optic tract* from the chiasm to the lateral geniculate and pulvinar. Other optic tract fibers can be traced over the surface of the thalamus to the superior colliculus. Grasp the optic chiasm between thumb and forefinger and lift gently to pull the fibers of the optic tract away from the surface of the thalamus. This method will allow you to follow the optic tract fibers to their terminations (see Figure 2.24). Visual fibers project from the lateral geniculate and pulvinar and travel to the occipital lobe of the cerebral cortex. These cannot be dissected very well in the sheep.

d. Auditory system

The *medial geniculate body* (thalamic auditory way-station) and the *inferior colliculus* (midbrain auditory reflex center) can be located on the lateral surface along with the *brachium of the inferior colliculus*. This structure forms the auditory pathway between the inferior colliculus and the medial geniculate (see Figure 2.24). Auditory projection fibers from the medial geniculate travel through the internal capsule to the cerebral cortex. These cannot be dissected very well in the sheep.

40

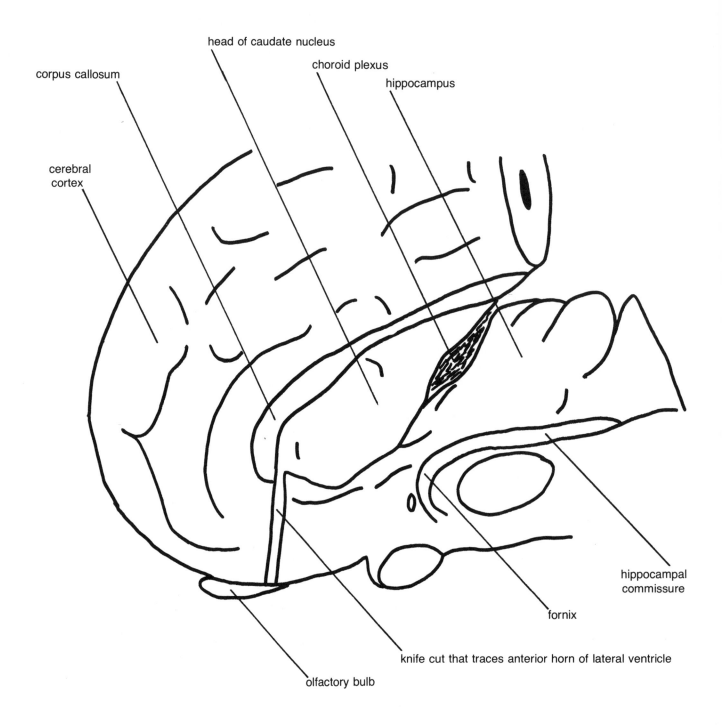

Figure 2.22. *Medial view of the anterior portion of the right cerebral hemisphere. The head of the caudate nucleus, choroid plexus of the lateral ventricle, and the hippocampus can be seen. Note also the knife cut which traces the anterior horn of the lateral ventricle.*

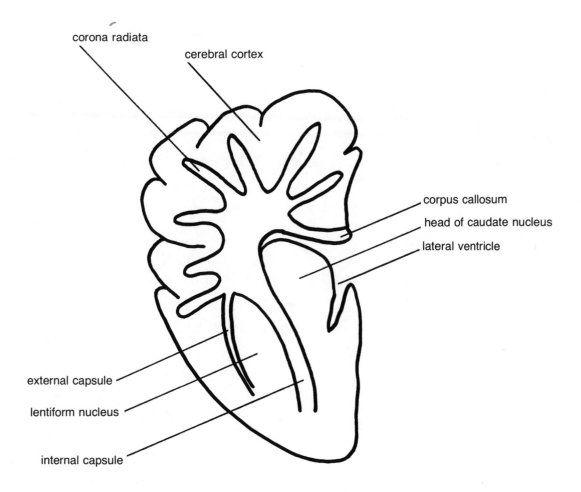

corona radiata

cerebral cortex

corpus callosum

head of caudate nucleus

lateral ventricle

external capsule

lentiform nucleus

internal capsule

Figure 2.23. *Frontal section through the right cerebral hemisphere at the level of the caudate nucleus, internal capsule, and lentiform nucleus.*

42

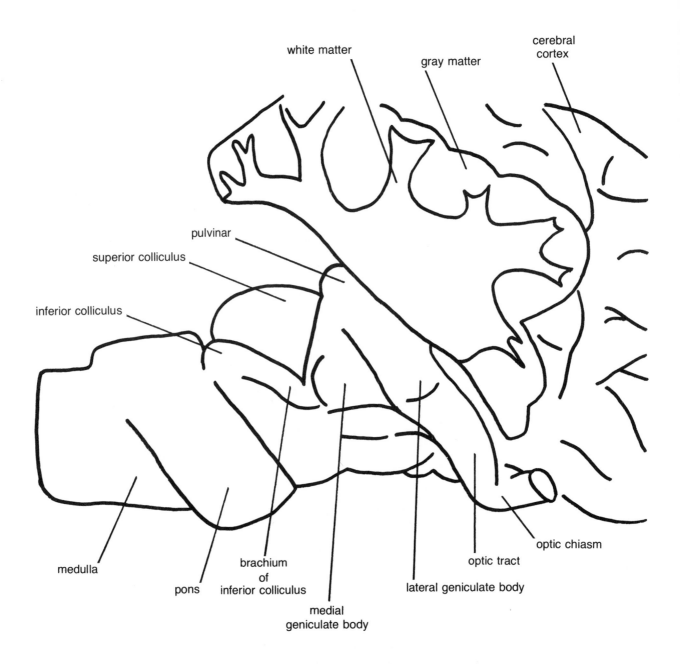

Figure 2.24. *Lateral view of the right half of the brain with the cerebellum, posterior pole of the cerebral cortex, and hippocampus removed. The major thalamic and midbrain structures of both the visual and the auditory systems can be seen.*

10. HORIZONTAL SECTIONING OF THE LEFT HEMISPHERE

Now cut the entire left hemisphere of the sheep brain into four or five horizontal sections. A long knife or a microtome blade is best suited for this purpose. A suggested method of making these sections is first to cut a "horizontal" section through the neocortex about 3 mm. dorsal to the corpus callosum and *in the same plane* as this structure. This section will actually slope ventrally toward the frontal cortex as does the corpus callosum. A second cut can be made just ventral to the body of the corpus callosum. Observe the relations of the choroid plexus of the lateral ventricle, the hippocampus, the fimbria, and the head of the caudate nucleus to one another and to the lateral ventricle.

Make another section in the same plane, but at a somewhat smaller angle from the horizontal section, through the hippocampus and the head of the caudate nucleus. Note the relations of the internal capsule to the adjacent structures. The *corpus striatum* is made up of the lentiform nucleus lying far laterally, the caudate nucleus lying medially, and the internal capsule, appearing as a band of white fibers between these two masses of gray matter, as well as between the lentiform nucleus and the thalamus.

Horizontal sections can then be made at several locations to expose the subcortical structures already studied. Look particularly for the continuation of the internal capsule into the cerebral peduncle (see Figures 2.25 and 2.26). Compare the structures seen in the horizontal sections with the tract and other dissections you have already carried out. Try to visualize a three-dimensional view of the fiber tracts of the brain.

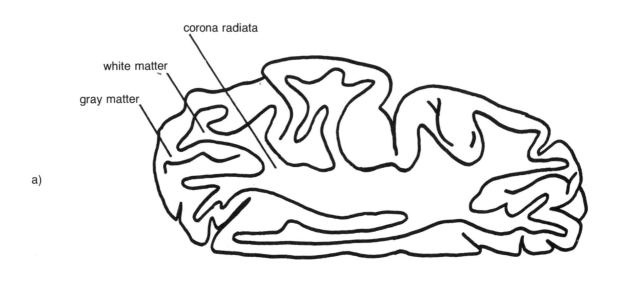

a)

b)

Figure 2.25. *A series of five horizontal sections through the left hemisphere progressing from the cerebral cortex ventrally to the thalamus and midbrain.*

45

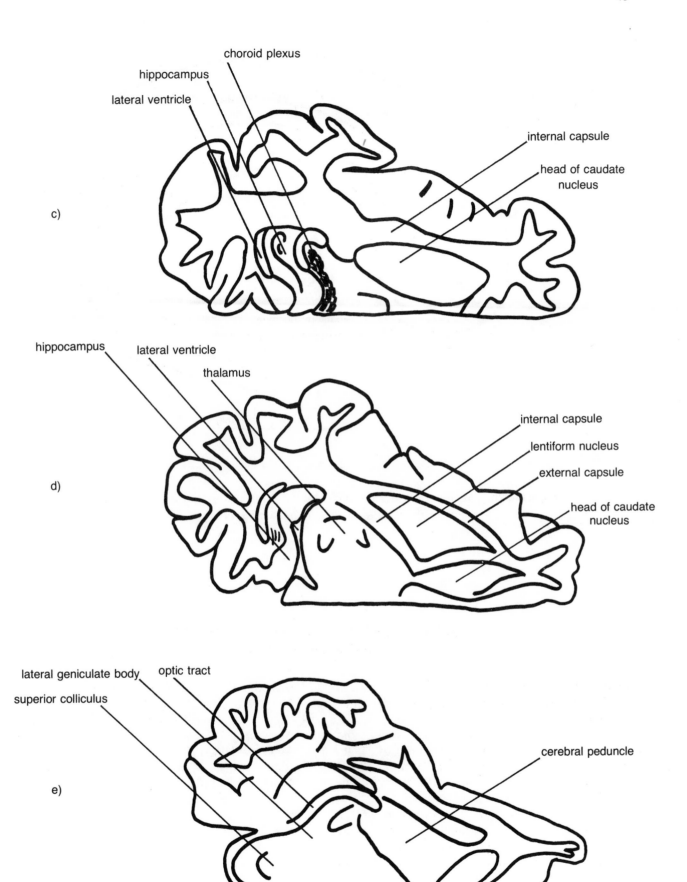

c)

choroid plexus
hippocampus
lateral ventricle

internal capsule

head of caudate
nucleus

d)

hippocampus
lateral ventricle
thalamus

internal capsule

lentiform nucleus

external capsule

head of caudate
nucleus

e)

lateral geniculate body   optic tract

superior colliculus

cerebral peduncle

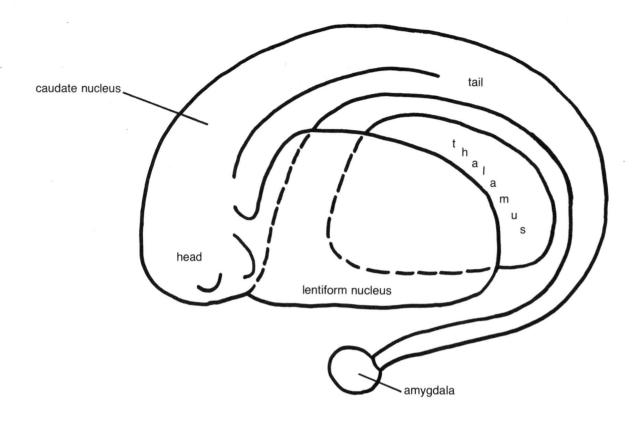

Figure 2.26. *Schematic diagram showing the relationship of the caudate nucleus to the lentiform nucleus and thalamus.*

REFERENCES AND SUGGESTIONS FOR FURTHER READING

Benson, H. J., and Gunstream, S. E. *Anatomy and Physiology Laboratory Manual, 5th Ed.* Dubuque, Ia.: Brown, 1974. pp. 72-85.

Briggs, E. A. *Anatomy of the Sheep's Brain, 2nd Ed.* London: Angus and Robertson, Ltd., 1946.

Fiske, E. W. *An Elementary Study of the Brain Based on the Dissection of the Brain of the Sheep.* New York: Macmillan, 1913.

Guyton, A. C. *Structure and Function of the Nervous System.* Philadelphia: Saunders, 1972.

May, N. D. S. *The Anatomy of the Sheep, 3rd Ed.* St. Lucia, Queensland, Australia: University of Queensland Press, 1970. pp. 236-266.

Netter, F. *The CIBA Collection of Medical Illus-trations: Volume 1–Nervous System.* Summit, N.J.: CIBA Pharmaceutical Co., 1962.

Noback, C. R., and Demarest, R. J. *The Human Nervous System, 2nd Ed.* New York: McGraw-Hill, 1975.

Northcutt, R. C., Williams, K. L., and Barber, R. P. *Atlas of the Sheep Brain, 2nd Ed.* Champaign, Ill.: Stipes, 1966.

Ranson, S. W., and Clark, S. L. *The Anatomy of the Nervous System, 10th Ed.* Philadelphia: Saunders, 1959.

Truex, R. C., and Carpenter, M. B. *Strong and Elwyn's Human Neuroanatomy, 5th Ed.* Baltimore: Williams and Wilkins, 1964.

3

Dissection of
the Sheep Eye

The sheep eye is a typical mammalian eye, but it has adaptations characteristic of grazing animals, particularly those which have been domesticated. The eyes are located well to the side of the head, providing a total field of vision of about 350° with a small binocular field of about 25° ahead. With this arrangement maximal alertness can be maintained for monocular vision, but not for binocular cues for objects situated at a distance. In man the frontal location of the eyes provides a total visual field of about 200° with a binocular overlap of about 110°. The eyes of the human are thus specialized by position for binocular depth perception of relatively close objects. Other differences between the sheep eye and the human eye will be noted below.

The sheep eye that has been provided for you has been fixed in a 10% formalin solution. Formalin causes fixation which arrests decomposition and hardens the tissues by coagulation of the proteins. The fixed eye is easier to handle and retains its form better. Fixation, however, causes color changes (muscles becoming light brown) and loss of transparency of the cornea and lens. The following instruments and equipment will be needed to carry out the dissection properly: scalpel, forceps, scissors, wooden orange stick, and two-inch-deep specimen dish. Laboratory work should be supplemented by study of textbooks. Blank pages are provided at the back of the manual so that you may record your observations and make drawings.

1. EXTERIOR OF THE GLOBE

In dissecting the eye it is important to note the aspect in which the structures are located (anterior, posterior, superior, inferior, lateral, medial, etc.). Examine the external appearance of the eye. Some of the structures which may remain intact on the globe are the *extrinsic muscles, orbital fat, root of the optic nerve, eyelids* including the third eyelid or *nictitating membrane,* and *conjunctiva.* Locate the positions of the eye with reference to the six aspects mentioned above. Some clues which may aid in this identification are the shape of the pupil and cornea, the position of the optic nerve, and the location of the nictitating membrane. The pupil when contracted is pear-shaped, with the broad end situated medially or nasally. On the posterior aspect of the globe the foot of the optic nerve is located somewhat ventral to the medial horizontal plane and it curves nasally. The nictitating membrane is located in the medial and ventral quadrant of the eye when viewed from the front. In the human a vestige of the nictitating membrane (the plica semilunaris) can be seen in the medial corner of the eye situated behind a pink pad of tissue (see Figure 3.1).

48

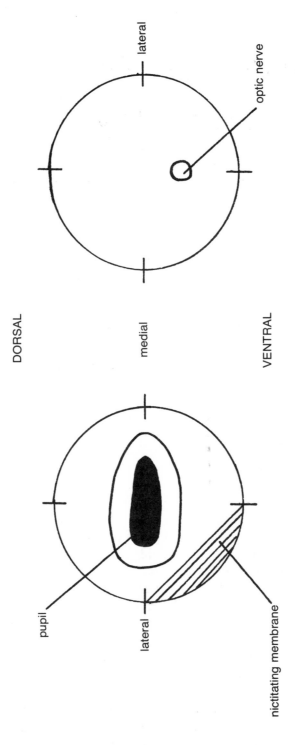

Figure 3.1. *Anterior and posterior views of the left eyeball. Note the locations of the nictitating membrane, pupillary enlargement, and optic nerve.*

The specific extrinsic muscles need not be identified by name on your specimen, since some of them have probably been torn away at the insertion. However, their functions will be briefly described. There are six muscles that control the orientation of the eye in space: (1) the *superior rectus,* (2) the *inferior rectus,* (3) the *lateral rectus,* (4) the *medial rectus,* (5) the *superior oblique,* and (6) the *inferior oblique.* The first two muscles are located on the antero-dorsal and ventral aspects of the globe respectively, and they rotate the eyeball upward and downward. The lateral and medial rectus are located on the antero-lateral and medial aspects of the globe, and they rotate the eyeball nasally and temporally. The superior and inferior oblique muscles are responsible for tilting of the eye. The superior oblique is attached to the dorso-lateral surface of the globe midway between the entrance of the optic nerve and the cornea. The inferior oblique passes laterally and curves upward and attaches to the lateral side of the globe under cover of the lateral rectus, close to the attachment of the superior oblique. Any rotation of the eye is due to an intricate coordination of all of these muscles responding at once (see Figure 3.2).

Three cranial nerves innervate these muscles—the oculomotor, the trochlear, and the abducens. There are two other muscles attached to the eye. The *levator* is located above the superior rectus and upon contraction it raises the upper eyelid. Both the levator and superior rectus are served by the oculomotor nerve and they act conjointly. Thus when the superior rectus contracts to rotate the eye upward, the levator also contracts to withdraw the upper eyelid from the field of vision. The other muscle referred to is the *retractor* which surrounds the optic nerve and which is inserted around the equator of the globe (the equator is defined with reference to the anterior and posterior poles). The retractor, absent in the human, pulls the entire eye deeper within the orbit. As a protective mechanism this action also results in the passive closure of the nictitating membrane over the sclera and cornea. The third eyelid has no muscle of its own, but is served indirectly by the retractor. The abducens nerve innervates the retractor. The lower eyelid has no separate muscle either, but a branch of the inferior rectus is attached to it, providing some degree of movement coordinated with oculorotation (see Figure 3.3).

The yellowish orbital fat in which the muscles are embedded functions as a cushion absorbing any mechanical shocks to the eye and it also supports the globe in the orbit.

Around the root of the optic nerve note the *dural sheath.* This sheath is continuous with the dura mater of the brain, consistent with the embryological development of the optic nerve and retina as an extension of the CNS.

The *conjunctiva* is a delicate folded membrane arising at the margin of the cornea and sclera (called the limbus region) which covers the sclera for a short distance before folding backward to line the inner surface of the eyelids. The conjunctiva secretes a watery fluid lubricating the area between the eyelids and the cornea.

Now with the aid of a scalpel scrape away the extrinsic muscles, orbital fat, and eyelids, but preserve the root of the optic nerve. These structures as well as others surrounding the eye are collectively termed the *ocular adnexa.* After the adnexa is removed, the sclera will be completely exposed over the entire surface of the globe. With a light pressure of the fingers and thumb note any differences in scleral thickness at different points on the globe. Is the globe spherical or is one axis relatively short? Note any pigmentation at the corneal margin. Make a few corneal abrasions by scraping its surface with a scalpel. If the cornea is not too cloudy, the position and shape of the pupil should be noted. How does the sheep pupil differ from the human pupil? The innervation of the cornea is standard in most mammals. Fine corneal nerve fibers emanate from the pericorneal plexus deriving from the ciliary nerves. These nerves are both sensory and motor in function, supplying fibers to the ciliary muscle, iris, and cornea, and they constitute two divisions of the ophthalmic branch of the trigeminal nerve. All of the cutaneous sensations can be elicited from the cornea.

Now bisect the globe around the equator. Hold the globe between the thumb and forefinger and make an incision through the sclera

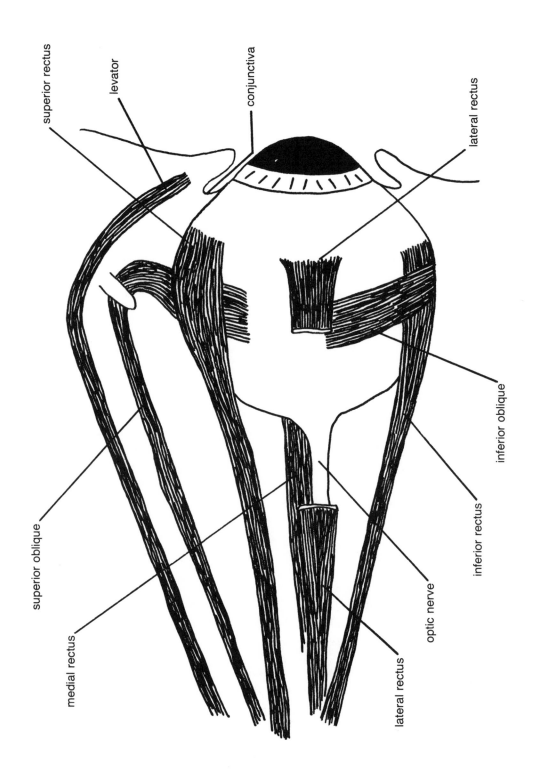

Figure 3.2. *Lateral view of the right eyeball showing the locations of six of the extrinsic eye muscles.*

51

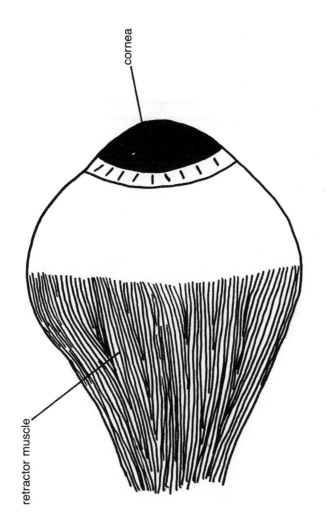

cornea

retractor muscle

Figure 3.3. *Dorsal view of the globe showing the location of the retractor muscle after the other extrinsic muscles have been removed.*

with the scalpel, using a sawing action. On penetration, registered by an escape of fluid, insert a pair of scissors and cut all around the equator. Without letting the two halves separate, take the scalpel and gently cut through the *vitreous humor* with a sawing action. Unless the vitreous is first cut it is liable to come away as a whole, bringing the lens and retina with it. This method of bisection avoids the lens which is very hard (see Figure 3.4).

2. INTERIOR OF THE GLOBE

Place the two halves in water to avoid drying and float out some of the more delicate structures. Note the inequalities in scleral thickness.

a. Posterior section

The appearance of the posterior half is similar to the view the ophthalmologist receives when examining the eye. This hemispherical bowl is called the *fundus.* Look for the *optic nerve head* or disc which is devoid of receptor elements (blind spot), the filmy *retina* and vessels (folds in the retina are artifacts), the *choroid* (black, middle layer of the eye), and any surface coloration it may have (the *tapetum*). Except in primate animals no central retinal area (or fovea) is macroscopically distinguishable.

In the sheep the receptors of the retina are mostly rods. The cones begin to manifest themselves in small numbers a little way from the periphery. As the cones become more numerous toward the center, the number of ganglion cells (optic nerve cells) also increases. This corresponds to the observation that cones tend to have "private-line" connections with bipolar cells (second-order neurons) and ganglion cells (third-order neurons), while rods have chiefly "party-line" connections.

The *choroid* is a heavily pigmented layer behind the retina that provides a dark chamber for the eye and it absorbs incident light before it can be reflected. On the inner surface of the choroid is a brilliant patch of colors, the *tapetum,* which varies from yellow to blue to

violet. It is responsible for the bright glow so readily seen in the pupil of certain animals at night when facing a source of light. The tapetum is not found in the human eye. Although its function is uncertain, it may act as a back reflector, increasing the amount of light passing through the retina, and hence its sensitivity at low levels of illumination. The receptor elements of the retina are located adjacent to the inner surface of the choroid with the bipolar cells, ganglion cells, and blood vessels lying over them. Light passing through the *dioptric apparatus* (cornea, aqueous humor, lens, and vitreous humor) must therefore pass through these retinal layers before stimulating the photoreceptors or being absorbed by the choroid. Gently pull a part of the retina away from the choroid and note its transparency (see Figure 3.5).

b. Anterior section

Now examine the anterior half of the globe. Note the sharp termination of the retina along a circle called the *ora ciliaris retinae* (ora serrata in man). Beyond this the choroid thickens to form the *ciliary body*. The ciliary body tapers down to form the *iris*. From the ciliary body emanate the *suspensory ligaments* of the *lens*. The deeply pigmented sphere including the choroid, ciliary body, and iris is termed the *uvea* or middle coat of the eye. A number of folds, the *ciliary processes,* can be easily distinguished on the inner surface of the ciliary body. The *lens* is located behind the iris.

After making these observations on the interior of the globe, the dissection of the two halves may begin. First it is necessary to remove the *vitreous,* a gel having the consistency of raw egg white. With brushing motions of the orange stick pull away the vitreous from the posterior half. If done properly, the retina and choroid will remain intact and in position. Place a small piece of the vitreous on a paper towel and notice the gradual loss of watery content (humor). The chamber between the lens and cornea is also filled by a watery fluid called the *aqueous humor.* The vitreous is similar to, if not identical with, the aqueous (see Figure 3.5).

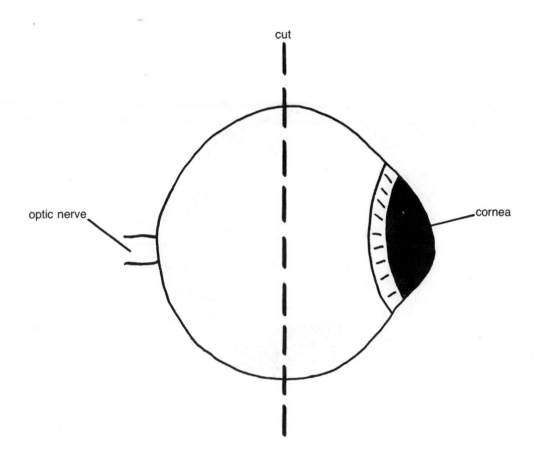

Figure 3.4. *Dorsal view of the globe showing the direction of cut for separating the eyeball into anterior and posterior sections.*

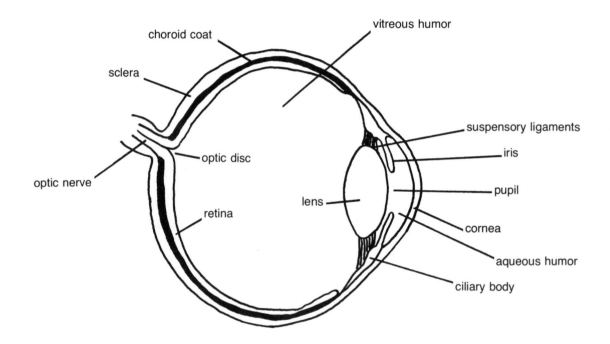

Figure 3.5. *Internal anatomy of the eye.*

3. DISSECTION OF THE POSTERIOR HALF

After removal of the vitreous replace the posterior half under water and note how the retina comes away readily from the underlying choroid but remains attached at the optic disc. The greater strength of the retina in this area is due to an increased density of neuroglial cells around the optic disc. As in the brain the glial cells of the retina are supportive and connective in function as well as serving to insulate one neuron from another. The blood vessels of the retina are visible only insofar as they retain some blood in them. They tend to follow the retinal folds.

With a scalpel bisect the posterior half through the optic nerve root and optic disc. Note that the nerve is narrowest where it traverses the *scleral foramen*. With blunt forceps strip away the retina and examine the choroid and tapetum. Scrape away some of the black pigment of the choroid near the margin of the tapetum. How far does the tapetum extend beneath this pigmented layer? Again with the forceps peel off the choroid. This exposes a pigmented layer on the inner surface of the sclera upon which the imprint of ciliary nerves and vessels shows as a light trace. As described above the ciliary nerves innervate the cornea, iris, and ciliary muscle. In the eye, many nerves and blood vessels follow similar routes located adjacent to one another. Many such analogous nerves and vessels are given corresponding names: thus ciliary nerves and ciliary vessels. The vascular system of the sheep eye differs considerably from that of the human eye. One of the major differences is that arterial influx and venous efflux are provided primarily by the external carotid and external jugular, respectively, rather than by their corresponding internal branches as in man.

4. DISSECTION OF THE ANTERIOR HALF

Invert the anterior half over a dish of water and then cut through the suspensory ligaments of the lens (attached to the ciliary body) by working the scalpel around and between the lens equator and the ciliary processes. Detach the lens and note any differences in curvature between the anterior and posterior surfaces. Then bisect the lens along the antero-posterior axis. Note the laminated structure and harder central nucleus. With your fingers break one of the lens halves into two quarters. The laminations may be more apparent now and a thin transparent membrane, the *lens capsule,* should be visible at the point of fracture. The suspensory ligaments are continuous with the lens capsule. The lens is the mechanism for ocular accommodation. When the ciliary muscle, located on the anterior aspect of the ciliary body, contracts in a sphincter-like fashion, tension on the suspensory ligaments is relieved, resulting in an increased curvature of the lens. Objects nearer the eye are thus brought into focus on the retina. The innervation for the accommodative reflex is not well understood. It is thought to be primarily a parasympathetic effect initiated by a change in the focus of light incident on the fovea. Pathways to the visual cortex and superior colliculus via the optic nerve and lateral geniculate bodies return to the ciliary muscle, most likely through the oculomotor nerve. In the sheep the ciliary muscle is very small and probably has little significance. In considering the role of this muscle in accommodation, ask yourself how its small size relates to the location of the eyes on the head.

Now inspect the inner part of the remaining portion of the anterior half. With blunt forceps test the attachment of the retina at the margin of the ciliary body. Remove any vitreous that may remain with an orange stick. Strip off the anterior uvea in one piece with forceps and place it aside. Examine the inner surface of the outer coat and then bisect it. At the edge of the cut look for differences in corneal thickness between the center and periphery. Note the overlap of the sclera over the cornea.

Now examine the anterior uvea. Note the white band marking the position of the ciliary body (and muscle), the iris pattern and pigment, and the pupil shape and size. In the sheep the iris is usually golden in color. When contracted, the pupil assumes a somewhat pear-shaped horizontal slit with the broader end situated medially in conformity with the shape of the cornea. However, upon dilation the pupil becomes more circular. Changes in pupil size are classic examples of the antagonism between the sympathetic and parasympathetic divisions of the autonomic nervous system. The dilator muscles of the iris, oriented radially and innervated by the sympathetic system, produce an increase in pupil size upon contraction. In opposite fashion the sphincter muscles of the iris, which are oriented concentrically and innervated by the parasympathetic system, cause the pupil to shrink upon contraction. The sphincters dominate the dilators. The dilators of the iris remain in relatively tonic contraction, exerting their influence when the sphincter reflex arc is inactive. Finally, on the upper and lower edges of the iris are located several deeply pigmented masses, the *corpora nigra*. These bodies are arranged in such a way that when the iris sphincters contract strongly the corpora nigra interlace to provide an additional degree of closure of the pupil beyond that supplied by the muscular response of the iris.

REFERENCES AND SUGGESTIONS FOR FURTHER READING

Benson, H. J., and Gunstream, S. E. *Anatomy and Physiology Laboratory Manual, 5th Ed.* Dubuque, Ia.: Brown, 1974. pp. 90-94.

Guyton, A. C. *Structure and Function of the Nervous System.* Philadelphia: Saunders, 1972. pp. 115-149.

Noback, C. R., and Demarest, R. J. *The Human Nervous System, 2nd Ed.* New York: McGraw-Hill, 1975. pp. 345-383.

Prince, J. H., Diesem, C. D., Eglitis, I., and Ruskell, G. L. *Anatomy and Histology of the Eye and Orbit in Domestic Animals.* Springfield, Ill.: Thomas, 1960. pp. 182-209.

Romanes, G. J. *Cunningham's Manual of Practical Anatomy, 13th Ed.* New York: Oxford University Press, 1967. pp. 67-68 and 196-203.

Spooner, J. D. *Ocular Anatomy.* London: The Hatton Press, Ltd., 1957. pp. 207-210.

4
Rat Brain
Serial Sections

The following pages contain outline drawings of the major structures of the rat brain shown in both frontal (coronal) and sagittal sections. A list of the abbreviations used to identify the various structures is also included. Fourteen frontal and four sagittal sections of one hemisphere, beginning in the anteriormost portions of the forebrain and continuing through the caudal portion of the midbrain, are illustrated (Figures 4.1 through 4.18).

These drawings are provided so that you may make comparisons between the rat brain and the sheep brain. Many of the structures previously located in the sheep brain can be identified on these rat brain sections. In addition, a few structures that could not be seen by gross dissection of the sheep brain have been identified in the rat brain sections. Note differences between the rat and sheep brains by comparing the sizes and shapes of various structures (e.g., many fewer convolutions in the cerebral cortex of the rat).

These serial maps or atlases of the rat brain are very helpful to researchers who are interested in locating specific brain structures in the living animal. Neuroanatomists who develop such maps or atlases of the brain use a device called a stereotaxic instrument which is designed with ear-bars and a mouth-piece so that the head of an anesthetized animal can be rigidly held in place. In addition, the stereotaxic instrument has an arm which can be moved with great precision in all three planes (anterior-posterior, medial-lateral, and dorsal-ventral). Accuracies of 0.1 millimeter (mm.) are often required. (See Chapter 5 for a further explanation of these procedures.) To develop such an atlas, serial sections of the brain are made with a microtome. Sections are usually 50-100 microns in thickness. Most brains are sectioned in the frontal plane, but sometimes they are sectioned in the sagittal plane. Both types of sections are shown in this chapter.

Both for sectioning and for accurately measuring distances within the brain a "zero" or reference point is needed. For sectioning and measuring in the frontal plane the *inter-aural line* (line that goes through the brain between the two ear canals) is frequently used. This line then becomes the "zero" reference and the location of any brain structure can be specified as the distance in millimeters anterior or posterior to the inter-aural line.

Another convenient reference used as a "zero" point is *bregma*—the point where the frontal and parietal bones of the rat's skull meet. Bregma is often used as a "zero" point during sterotaxic surgery. (See Chapter 5 for further details).

The sections illustrated in this chapter have been enlarged approximately 12.5 times above their actual size in order to make the identification of structures easier. The vertical lines drawn on Figure 4.2 indicate the level at which certain of the sagittal sections were made (Figures 4.15 and 4.17). Likewise, the vertical lines drawn on Figure 4.15 indicate the level at which certain of the frontal sections were made (Figures 4.1, 4.4, 4.9 and 4.13).

On each of the following figures the number in the legend indicates the distance that the section is anterior to the inter-aural line (e.g., A 10.3 mm.) for the frontal sections, or the distance that the section is lateral to the midline (e.g., L 0.2 mm.) for the sagittal sections.

LIST OF ABBREVIATIONS

A
ac=anterior commissure
amy=amygdala

B
bic=brachium of inferior colliculus

C
ca=cerebral aqueduct
cc=corpus callosum
cg=cingulum
cn=caudate nucleus
cp=cerebral peduncle
cx=cerebral cortex

D
dscp=decussation of superior cerebellar peduncle

E
ec=external capsule

F
fm=fimbria fibers
fx=fornix

H
h=habenula
hpc=hippocampus
hpcc=hippocampal commissure
hpt=habenulo-peduncular tract

I
ic=inferior colliculus
inc=internal capsule
ipn=interpeduncular nucleus

L
lgb=lateral geniculate body

lh=lateral hypothalamus
ln=lentiform nucleus
los=lateral olfactory stria
lv=lateral ventricle

M
mb=mammillary body
mfb=medial forebrain bundle
mgb=medial geniculate body
mtt=mammillothalamic tract

O
oc=optic chiasm
ot=optic tract

P
p=pons
pc=posterior commissure
ps=pituitary stalk
pv=pulvinar

R
rf=rhinal fissure

S
sa=septal area
scp=superior cerebellar peduncle
sc=superior colliculus
sm=stria medullaris
st=stria terminalis

V
vmh=ventromedial hypothalamus

II=optic nerve
III=oculomotor nerve
IIIv=third ventricle
V=trigeminal nerve

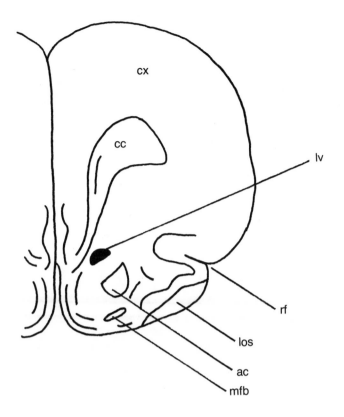

Figure 4.1. *A 10.3 mm.*

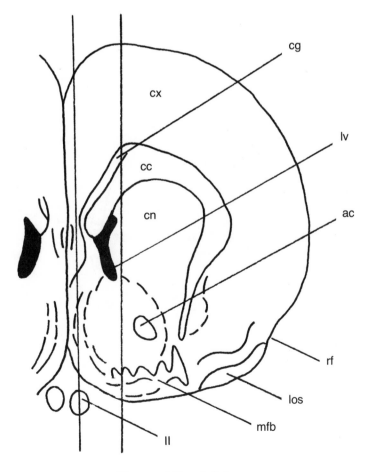

Figure 4.2. *A 9.4 mm.*

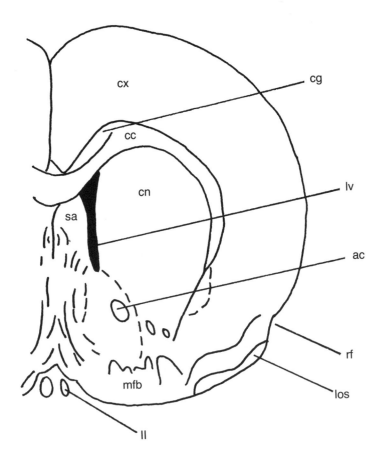

Figure 4.3. *A 8.6 mm.*

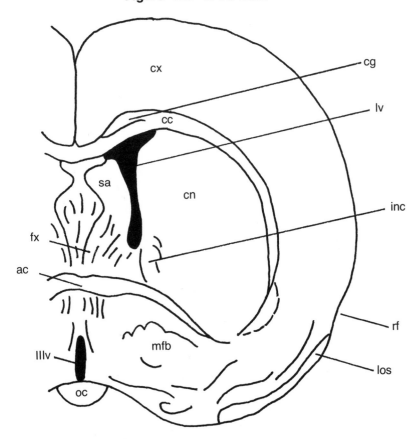

Figure 4.4. *A 7.2 mm.*

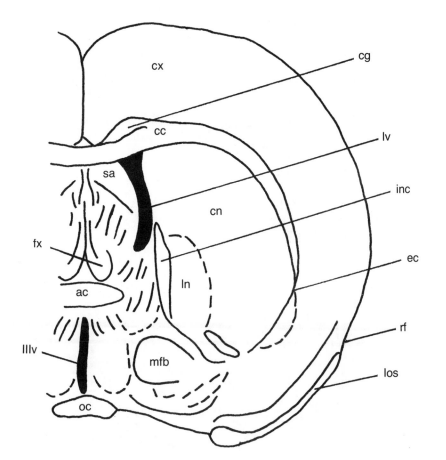

Figure 4.5. *A 6.9 mm.*

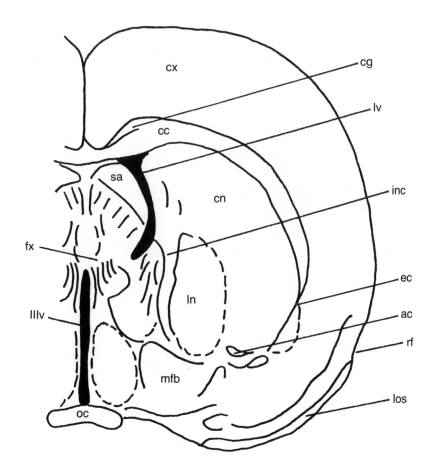

Figure 4.6. *A 6.7 mm.*

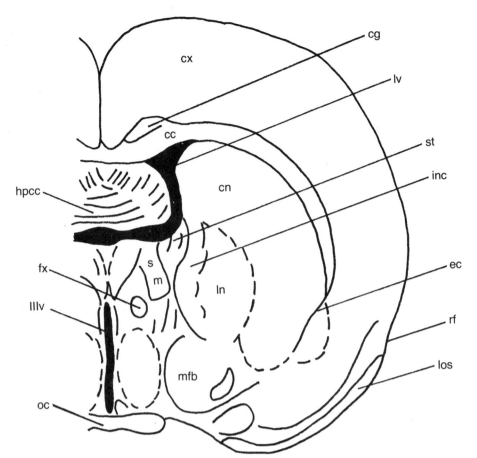

Figure 4.7. *A 6.4 mm.*

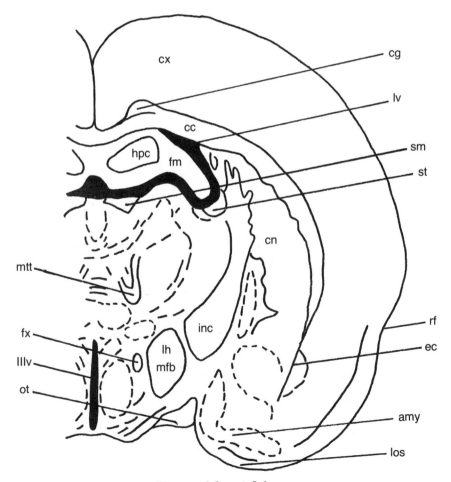

Figure 4.8. *A 5.3 mm.*

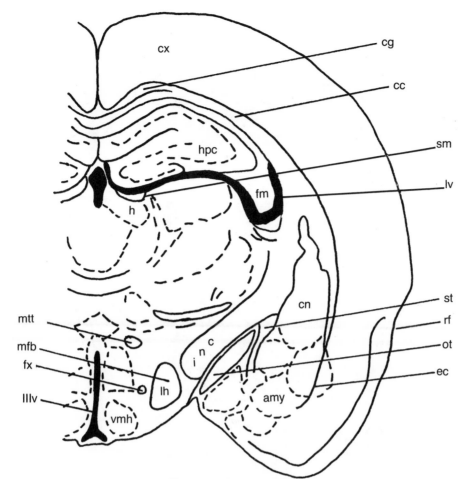

Figure 4.9. *A 4.1 mm.*

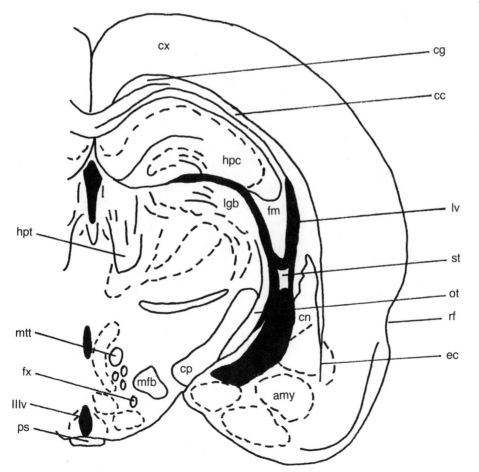

Figure 4.10. *A 3.4 mm.*

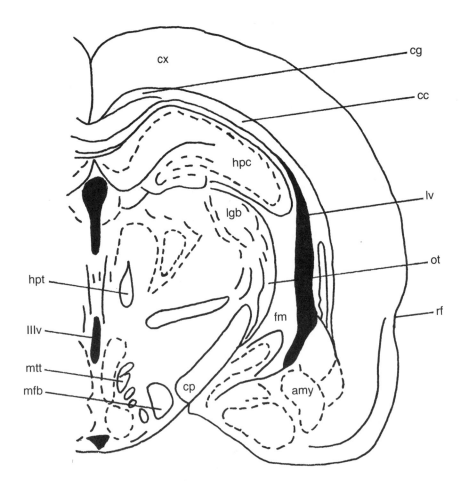

Figure 4.11. *A 3.3 mm.*

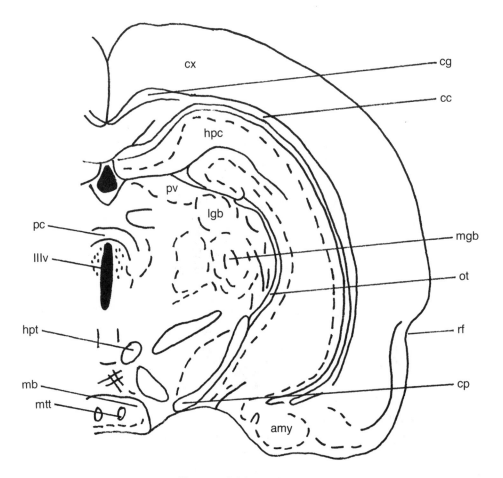

Figure 4.12. *A 2.6 mm.*

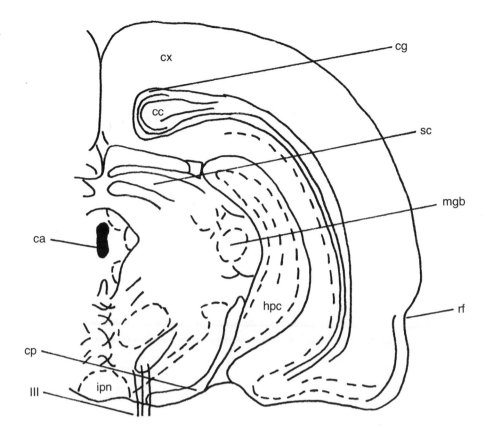

Figure 4.13. *A 1.8 mm.*

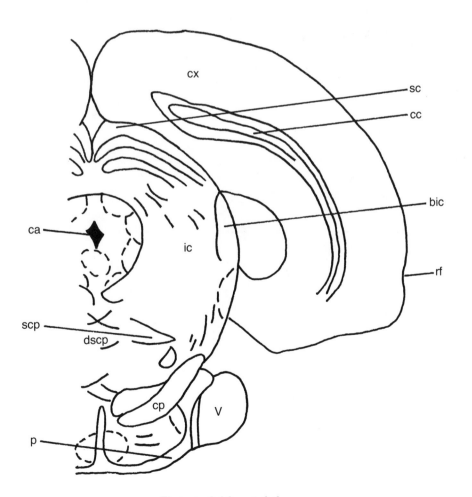

Figure 4.14. *A 0.6 mm.*

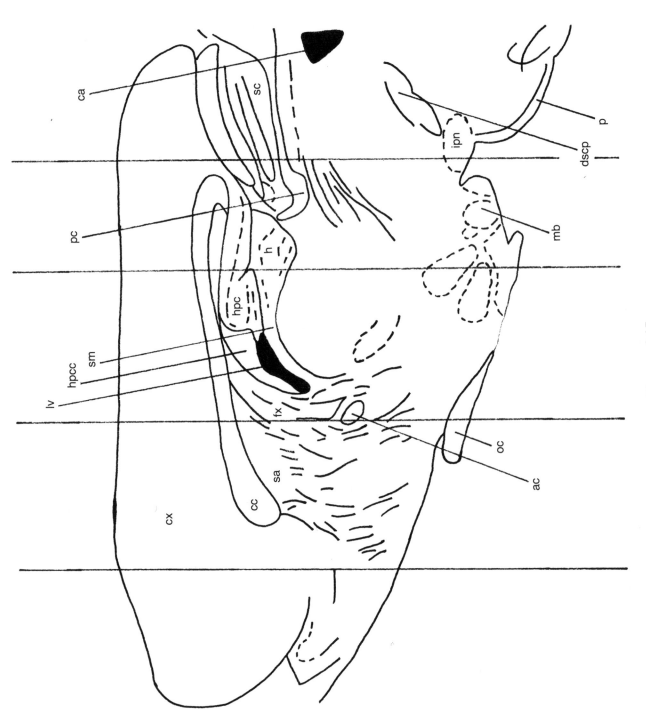

Figure 4.15.
L 0.2 mm.

67

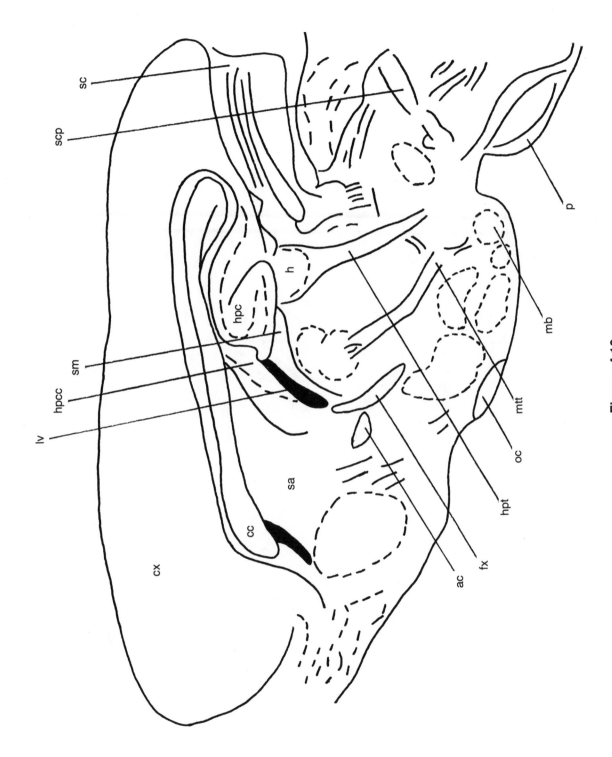

Figure 4.16.
L 0.7 mm.

68

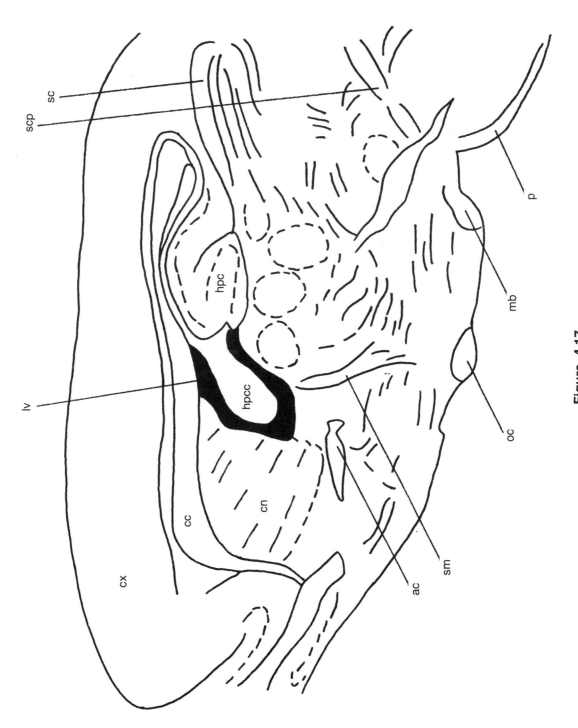

Figure 4.17.
L 1.2 mm.

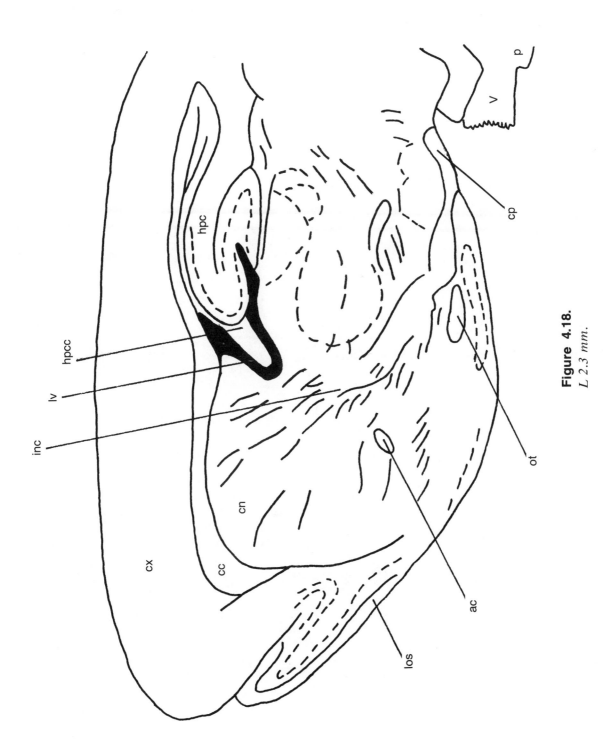

Figure 4.18.
L 2.3 mm.

REFERENCES AND SUGGESTIONS FOR
FURTHER READING

Hart, B. L. *Experimental Neuropsychology: A Laboratory Manual.* San Francisco: Freeman, 1969. pp. 94-102.

König, J. F. R., and Klippel, R. A. *The Rat Brain.* Huntington, N.Y.: Krieger, 1970.

Skinner, J. E. *Neuroscience: A Laboratory Manual.* Philadelphia: Saunders, 1971, pp. 87-143.

5

Stereotaxic Brain Surgery on the Rat

Much of the research in physiological psychology necessitates locating structures deep within the brain with great precision. As mentioned in Chapter 4, a device called a stereotaxic instrument is used when this type of precision is required during experimental procedures. The instrument is designed with ear-bars and a mouthpiece and nose clamp so that the head of an anesthetized animal can be rigidly held in place. In addition, the instrument has an electrode-carrying arm which can be moved with great precision in all three planes. Accuracies of 0.1 mm. are often required. To locate a structure precisely deep within the brain a "zero" or reference point is needed from which measurements in all three planes can be made. This chapter outlines a step-by-step procedure that can be followed in performing stereotaxic brain surgery on the rat. For this example the brain structure to be located will be the ventromedial hypothalamus (VMH). A bilateral electrolytic lesion is to be made in the VMH.

The anesthetization of the rat is the first step in the process of preparing for stereotaxic surgery. Most general anesthetics are administered in doses that are related to the animal's body weight. Thus, the animal's weight must be known in order to select the proper dosage of anesthetic. The most frequently used general anesthetic is *sodium pentobarbital* which is injected with a hypodermic syringe into the peritoneal (gut) cavity. Ether can also be used as a general anesthetic, but it is less preferred because it is more difficult to administer in exact doses and because it is so volatile. A general anesthetic is administered to the animal for two purposes: first, to prevent the animal from experiencing pain during the surgery; and second, to eliminate all movements of the animal by making it unconscious. After the animal is unconscious, other drugs are often administered. *Atropine sulfate,* an anticholinergic drug, can be administered to inhibit salivation that might interfere with the animal's respiration during surgery. This is given through an intraperitoneal injection. Additionally, *procaine hydrochloride,* a local anesthetic, can be injected under the scalp to eliminate sensory feedback from this area. The scalp of the rat is highly innervated with pain receptors, stimulation of which may lead to reflexive movements during surgery. A local anesthetic will eliminate this problem. After anesthetization the animal's head is shaved with electric clippers. This is done to provide a clearer and cleaner area in which to work.

The animal must now be mounted in the stereotaxic instrument. This is a critical step in the procedure since misalignment of the animal's head will lead to inaccurate measurements in locating the specified brain structure. The animal must first be centered in the instrument's ear-bars. These bars are firmly placed into the animal's auditory canals and locked into position. Calibrations on the bars allow them to be centered so that the animal's head is directly in the middle of the instrument. Now the animal's upper front incisor teeth are placed over the mouth bar and a nose clamp is brought down and firmly locked in place. This is done to stabilize the animal's head even more during the surgical procedures.

The surgical procedures can now be initiated. The animal's scalp is opened by making a ¾-inch midline incision with a scalpel. The scalp is then retracted and the skull bones are cleaned so that *bregma* (the point where the frontal and parietal bones of the skull meet) can be accurately located (see Figure 5.1). The adjustable stereotaxic arm and electrode carrier can now be moved into place directly over bregma so that the tip of the electrode is just touching the bregma point. "Zero" readings are made for all three planes (anterior-posterior, medial-lateral, and dorsal-ventral) so that bregma is located in three-dimensional space. The next step is to move the electrode through a set of predetermined coordinates that will locate the VMH in relation to bregma. As mentioned in Chapter 4, these coordinates can be determined through the use of a stereotaxic atlas of the rat brain. The pages of the atlas contain serial sections of the rat brain that are ordered from anterior to posterior. A page is located that contains the appropriate brain structure (in this case the VMH). This page will have a measurement on it that specifies the number of millimeters that this section is anterior (or posterior) to the bregma point. In addition, each page is calibrated with horizontal and vertical measurements so that the brain structure's lateral distance from the midline and also its depth from the surface of the brain can be determined. In order to verify the accuracy of the chosen coordinates, a few pilot subjects may be operated on before the actual experiment begins. Using the above procedures the following coordinates for the VMH were determined: 2.7 mm. *posterior* to bregma, 0.5 mm. *lateral* to the midline, and 9.7 mm. *below* the surface of the skull (see Figure 5.2). To account for the distance between the surface of the brain and the outer surface of the

skull, 1.0 mm. is usually added to the depth coordinate taken from the atlas.

To locate the VMH the electrode tip is moved the specified distances from bregma. First, the electrode is moved 2.7 mm. posterior to bregma and then it is moved 0.5 mm. lateral from the midline. The point on the skull directly under the electrode tip is now marked and the movable stereotaxic arm is swung out of the way. Using a dental drill a small hole is made in the animal's skull. Care must be taken not to allow the drill tip to penetrate the underlying dura or brain tissue. After the skull hole is drilled, a fine needle is gently inserted into the opening to pierce the dura mater. This is done because the dura is so tough that it might deflect the course of the penetrating electrode if it were allowed to remain intact. Once the dura is pierced, the stereotaxic arm can be repositioned and the electrode lowered into the brain to a depth of 9.7 mm.

The lesioning electrode is connected to the lesion-making device. The circuit should be connected so that the lesioning electrode within the brain becomes the anode. A rectal or tail cathode can be used to complete the circuit through the animal. The cathode should have a large surface area so that the current density will be low. This will prevent the occurrence of a lesion at this part of the circuit. After the circuit has been connected, a predetermined amount of direct current (DC) is allowed to pass through the circuit for a specified time. Varying the current and time parameters will produce variations in the sizes and shapes of the lesions. Appropriate current and time parameters must be experimentally determined through the use of pilot animals. To damage the VMH extensively a current of 1.0 milliampere (mA) passed for 10 seconds has been found to be effective.

After the current has been passed, the lesion electrode is withdrawn from the brain and the same procedure is carried out on the opposite side of the brain to make the lesion bilateral. Following the second lesion and withdrawal of the electrode, the animal's scalp is sutured. The animal is then removed from the stereotaxic instrument and returned to its home cage. An intramuscular injection of

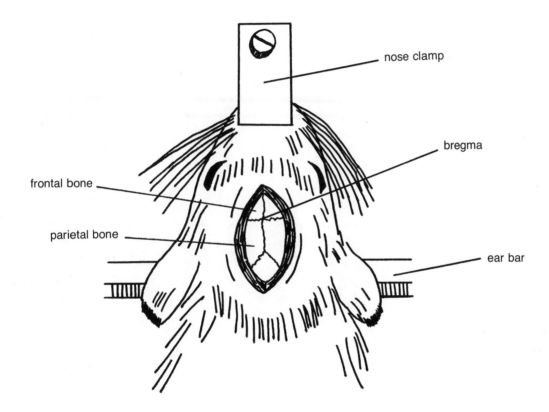

Figure 5.1. *Dorsal view of rat's head positioned in stereotaxic instrument with ear-bars and nose clamp. The scalp has been retracted to reveal bregma, the point where the frontal and parietal bones come together.*

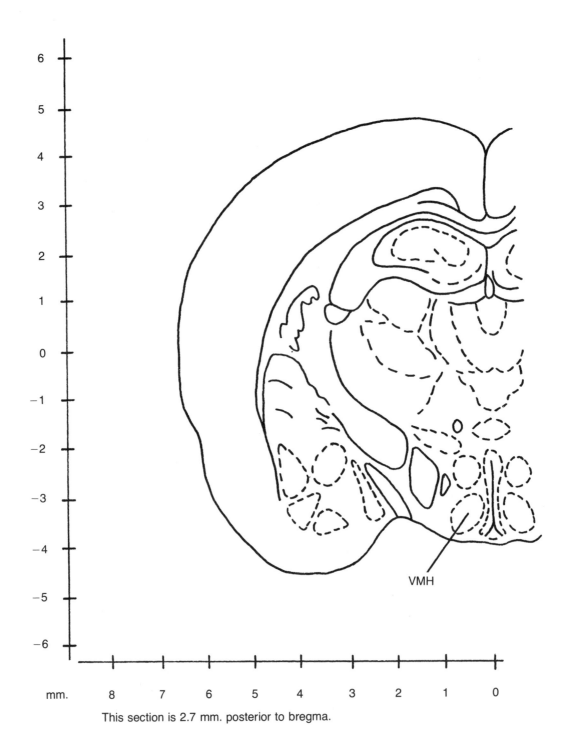

6
5
4
3
2
1
0
−1
−2
−3
−4
−5
−6

VMH

mm.   8   7   6   5   4   3   2   1   0

This section is 2.7 mm. posterior to bregma.

Figure 5.2. *Frontal section through rat brain at the level of the ventromedial hypothalamus (VMH). The location of the VMH in all three planes can be ascertained by taking the appropriate measurements from this figure.*

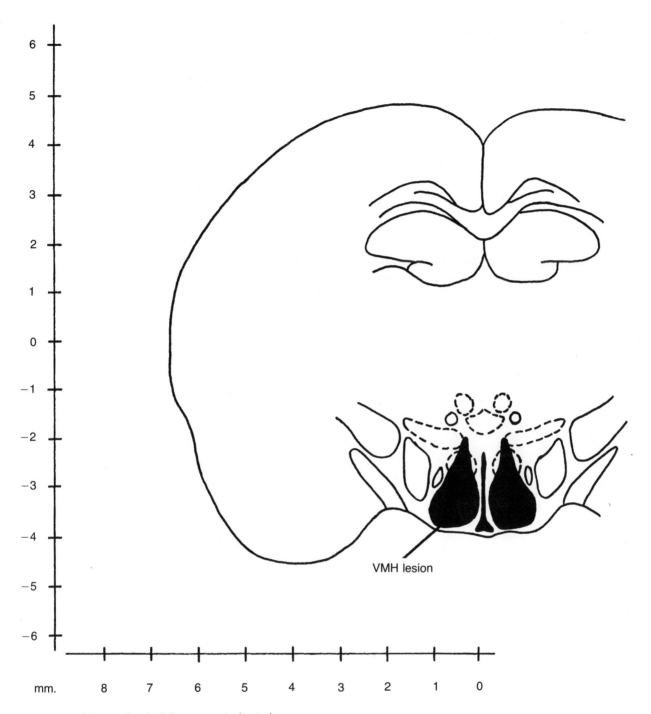

This section is 2.7 mm. posterior to bregma.

Figure 5.3. *Reconstruction of a bilateral VMH electrolytic lesion at the same frontal section of the brain as shown in Figure 5.2. Such reconstructions should be made for the entire anterior-posterior extent of the lesion.*

penicillin may be given at this time to guard against infection. During recovery in the home cage, the animal should be placed on a paper towel to help it overcome the hypothermia induced by the sodium pentobarbital.

The recording of behavioral data may commence as soon as the animal has recovered from surgery. When this phase of the experiment has been completed the animal must be sacrificed so that a histological examination of its brain can be made. This is necessary to verify the locus and size of the lesions. The animal is first given an overdose of sodium pentobarbital and then perfused through the heart with a physiological saline solution (0.9% NaCl) followed by a 10% formalin solution. After the perfusion is complete, the animal's head is re-moved and stored in formalin for approximately one week. The brain can then be carefully removed from the skull case with the use of a pair of rongeurs. The brain is either embedded in paraffin or frozen and sectioned frontally with a microtome. Sections are usually about 50 microns thick. The sections are mounted on microscope slides and are either stained with various dyes or photographed. The locus and size of the lesion can now be verified. Reconstructions of the lesions are often made in order that a permanent record can be kept (see Figure 5.3). The entire extent of the lesions should be recorded. Animals with lesions that missed the intended area or that heavily damaged adjacent tissue should be discarded from an experiment.

REFERENCES AND SUGGESTIONS FOR FURTHER READING

Hart, B. L. *Experimental Neuropsychology: A Laboratory Manual.* San Francisco: Freeman, 1969. pp. 22-44.

Skinner, J. E. *Neuroscience: A Laboratory Manual.* Philadelphia: Saunders, 1971. pp. 84-143.

6

Functional Neuroanatomy

The results of numerous stimulation and lesion studies have tended to support the notion that there are specific functions for specific brain areas. Yet it is also important to keep in mind that the nervous system possesses a remarkable capacity for recovering functions that were lost owing to some type of organic damage. In addition, it has frequently been observed that a rather large amount of tissue damage can be sustained without a noticeable behavioral impairment.

Neurologists are physicians who are specifically trained to detect and locate organic damage to the nervous system. To the clinical neurologist a patient represents the problem of interpretation of variations from an accepted normal standard. Such an interpretation requires knowledge of what is normal and what is abnormal. As part of a routine physical examination or in response to a specific complaint from a patient, a physician will test the functioning of various parts of the nervous system. The patient's responses to specific environmental stimuli will be tested. If a more thorough *neurological examination* is given, then all the patient's cranial and spinal nerves will be tested and a check for abnormal function will be made. Further testing may reveal the functional state of more centrally located portions of the nervous system from the reflex centers in the spinal cord and brainstem to the cerebellum and cerebral cortex. Other specific tests are designed to examine the patient's speech, state of consciousness, memory, and intelligence. Deficits can be sensory, motor, or sensory and motor, or they can involve the more complex higher functions. The examiner's task is to locate with as much precision as possible the source of any neurological deficit that is revealed.

Listed below are a number of cases in which the individual is experiencing a deficit in the functioning of part or parts of the nervous system. The behavioral impairments each patient is experiencing are given. Using your knowledge of neuroanatomy and structural-functional relationships, attempt to identify the parts of the nervous system directly involved in each of the behavioral dysfunctions. Knowledge of cranial nerve function as well as the function of various brainstem and cerebral areas will be necessary to determine correctly the locus of all the deficits.

Case 1

A sixty-two–year–old man was experiencing double or blurred vision, and he was having difficulty maintaining his balance and orientation in space. A neurological examination also revealed a slight deficit in the man's ability to control the muscles on the *left* side of his face. Where must the lesion be located?

Case 2

A forty-five–year–old man was experiencing a slight motor dysfunction on the *right* side of his body. In addition, he was having difficulty in focusing with his *left* eye. Where must the lesion be located?

Case 3

A twenty-seven–year–old woman complained of a partial loss of sight. A neurological examination revealed the following visual field

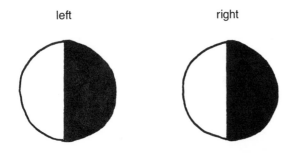

left   right

deficit. (The shaded portions represent areas of loss of vision. They do *not* indicate what part of the visual apparatus is not functioning, but what the person cannot see.) In addition, it was discovered that the patient had a hearing deficit. Where must the lesion be located?

Case 4

A thirty-five–year–old man had been experiencing a loss of appetite for the last three months during which time he had lost 25 lb. In addition, a neurological examination indicated a slight motor dysfunction on the patient's *right* side. Where must the lesion be located?

Case 5

A seventy-three–year–old woman suddenly began to have severe losses of recent memory. For example, she could not remember in the afternoon what she had had for breakfast that day. However, she had a good ability to recall experiences from her past (e.g., childhood experiences). Where must the lesion be located?

Case 6

A fifty-six–year–old woman had suddenly become very difficult to arouse. That is, prolonged and intense stimulation was necessary to keep her in a conscious waking state. A neurological examination indicated that her basic sensory systems were functioning adequately. Where must the lesion be located?

Case 7

After being in an automobile accident, a twenty-one–year–old man had the following symptoms:
  a) inability to read or comprehend spoken language
  b) a motor deficit on the *right* side of his body
Where must the lesion be located?

Case 8

A thirty-nine–year–old man was referred for a neurological examination because he had developed periodic seizures which culminated in violence and attack behavior. What was the most probable locus of the neural dysfunction?

Case 9

An eighty-three-year–old man suffered an apparent stroke and was unconscious for several hours. After recovering consciousness, he had difficulty speaking because he was unable to move his tongue correctly. A neurological examination revealed that he had paralysis in his *right* arm and leg and that the musculature of the *left* side of his tongue lacked the proper tonus. Where must the dysfunction be located?

Refer to Figure 6.1 while reading the following. Fibers in the optic system (optic nerve, optic chiasm, optic tract) have a topographical organization. Objects from the medial (nasal) part of the visual field strike the lateral (temporal) part of the retina. Optic nerve fibers

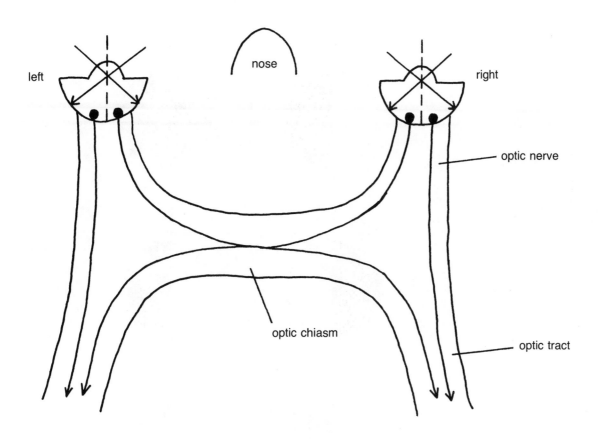

Figure 6.1. *Schematic diagram of the pathway followed by ganglion cell axons as they exit from various portions of the retina.*

from the lateral part of the retina do *not* cross at the optic chiasm, but instead continue into the optic tract on the *same* side of the brain. Objects from the lateral (temporal) part of the visual field strike the medial (nasal) part of the retina. Optic nerve fibers from the medial part of the retina *cross* at the optic chiasm and continue into the optic tract on the *opposite* side of the brain.

Lesions in various parts of the visual system can cause a variety of *visual field deficits*. Below are three such examples. The shaded portions represent areas of loss of vision. They do *not* represent what part of the visual apparatus is not functioning, but what the person cannot see. (R=right eye; L=left eye.)

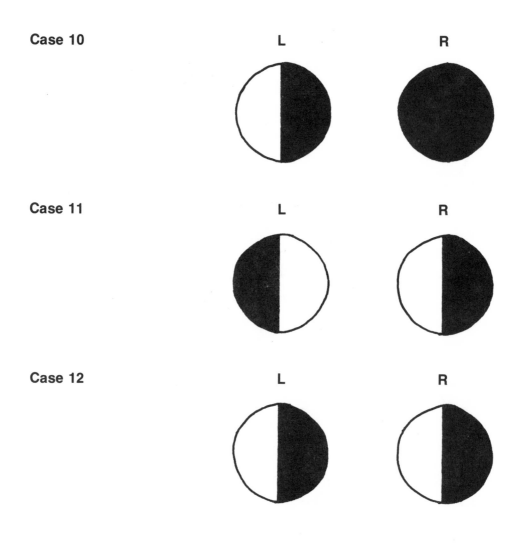

Case 10   L   R

Case 11   L   R

Case 12   L   R

REFERENCES AND SUGGESTIONS FOR FURTHER READING

Pincus, J. H., and Tucker, G. *Behavioral Neurology*. New York: Oxford University Press, 1974.

Ransom, S. W., and Clark, S. L. *The Anatomy of the Nervous System, 10th Ed.* Philadelphia: Saunders, 1959. pp. 419-446.

Appendix A

Glossary of Neuroanatomical Terms

Below is an alphabetical list of many of the neuroanatomical terms used throughout this laboratory manual. The derivation or origin of the terms is given so that you may gain a more complete understanding of them. (Source: Pepper, O.H.P. *Medical Etymology*. Philadelphia: Saunders, 1949. pp. 45-49.)

amygdaloid—Gr. amygdale=almond and eidos= resemble. The name given to the almond-shaped nucleus.

autonomic—Gr. auto=self and nomos=law. The part of the nervous system which is self-controlled or autonomous.

axon—Gr. axon=axis. Adopted for the name of the axis cylinder.

brachium—L. an arm or arm-like process.

brain—Gr. brechmos=forehead.

callosum—L. callous. Applied to the corpus callosum.

cerebellum—L. diminutive of cerebrum=brain.

cerebrum—L. brain.

cingulum—L. a girdle.

commissure—L. con (com)=together and mittere=to put. A joining or a seam.

convolution—L. con=together and volvere=to roll.

decussation—L. decussare=to intersect, from decussis=ten, represented by the symbol X, hence any crossing. The analogous word "chiasm" is a Greek derivation.

dendrite—Gr. pertaining to a tree. dendron=tree. The term dendrite is used of the processes of a nerve cell.

diencephalon—Gr. dia=through; and encephalon. Hence the "between" brain.

dura—L. durus=hard, mater=mother. The strong mother of the brain.

encephalon—Gr. enkephalon=brain, from en=in and kephale=head.

fasciculus—L. diminutive of fascis=bundle or packet.

ganglion—Gr. a swelling. Applied to a group of nerve cells.

geniculate—L. geniculare=to bend the knee; genu=knee.

glia—Gr. glue. A contraction used as a synonym for neuroglia.

gyrus—Gr. gyros=a circle. The words gyrate and gyroscope come from this word.

hippocampus—Gr. hippos=horse and kamos= monster. The curved gyrus which bears this name was so called because its shape suggests that of the sea horse.

hypophysis—Gr. hypo=under and physis=growth. Another name for the pituitary gland.

lemniscus—Gr. lemniskos=a band of fibers.

medulla—L. marrow. As in the medulla oblongata of the lower brainstem.

meninges—Gr. menix=membrane. The term meninges is reserved for the membranes covering the brain and spinal cord.

mesencephalon—Gr. meso=middle; and encephalon (see definition above).

myelin—Gr. myelos=marrow.

neuroglia—Gr. neuron=nerve and glia=glue.

neuron—Gr. nerve.

oblongata—L. oblongus=rather long or oblong.

parasympathetic—Gr. para=beside; and sympathetic. A name for a part of the autonomic nervous system.

pellucidum—L. per=through and lucere=to shine. Used of the septum pellucidum, through which light can shine.

pia—L. pius=kindly or tender. The tender protector of the brain and spinal cord.

pineal—L. pinea=pine cone. Named from the shape of this structure.

pituitary—L. pituita=mucous secretion. The mucous from the nose and mouth was once thought to come from the brain, hence this structure was so named.

plexus—L. something woven.

pons—L. a bridge. As in "pontoon."

pulvinar—L. a pillow. A part of the thalamus.

rhombencephalon—Gr. rhombos=a rhomb or four-sided structure; and encephalon.

splenium—Gr. splenion=bandage. Applied to any structure whose shape suggests a bandage.

striatum—L. straitus=furrowed. Applied to the corpus striatum (basal ganglia).

subcortical—L. sub=under and cortex=bark or outer covering. Applied to anything beneath the cortex of the brain.

sympathetic—Gr. syn=with and pathos=suffering.

tapetum—L. a tapestry or carpet.

tectum—L. a roof. The roof of the midbrain.

tegmentum—L. a cover. The upper covering of the cerebral peduncles.

telencephalon—Gr. telos=end; and encephalon.

thalamus—Gr. thalamos=an inner chamber. The anterior portion of the brainstem.

velum—L. a veil or covering.

Appendix B

List of Textbooks
of Neuropsychology

Below is a list of books which may be of general use in the study of neuropsychology.

Beatty, J. *Introduction to Physiological Psychology: Information Processing in the Nervous System.* Monterey, Calif.: Brooks/Cole, 1975.

Butter, C. M. *Neuropsychology: The Study of Brain and Behavior.* Monterey, Calif.: Brooks/Cole, 1968.

Deutsch, J. A., and Deutsch, D. *Physiological Psychology.* Homewood, Ill.: Dorsey Press, 1973.

Eccles, J. C. *The Understanding of the Brain.* New York: McGraw-Hill, 1973.

Gardiner, E. *Fundamentals of Neurology, 6th Ed.* Philadelphia: Saunders, 1975.

Grossman, S. P. *Essentials of Physiological Psychology.* New York: Wiley, 1973.

Grossman, S. P. *A Textbook of Physiological Psychology.* New York: Wiley, 1967.

Hubbard, J. I. *The Biological Basis of Mental Activity.* Reading, Mass.: Addison-Wesley, 1975.

Isaacson, R. L., Douglas, R. J., Lubar, J. F., and Schmaltz, L. W. *A Primer of Physiological Psychology.* New York: Harper and Row, 1971.

Leukel, F. *Introduction to Physiological Psychology, 3rd Ed.* St. Louis: Mosby, 1976.

Milner, P. M. *Physiological Psychology.* New York: Holt, Rinehart, and Winston, 1970.

Morgan, C. T. *Physiological Psychology, 3rd Ed.* New York: McGraw-Hill, 1965.

Schneider, A. M., and Tarshis, B. *An Introduction to Physiological Psychology.* New York: Random House, 1975.

Schwartz, M. *Physiological Psychology.* New York: Appleton-Century-Crofts, 1973.

Skinner, J. E. *Neuroscience: A Laboratory Manual.* Philadelphia: Saunders, 1971.

Teyler, T. J. *A Primer of Psychobiology.* San Francisco: Freeman, 1975.

Thompson, R. F. *Introduction to Physiological Psychology.* New York: Harper and Row, 1975.

Thompson, R. F. *Introduction to Biopsychology.* San Francisco: Albion, 1973.

NOTES AND DRAWINGS